HOW TO BUILD YOUR BRAND WITH A BOOK

Establishing Yourself as the Published Expert

SCOTT TURMAN
ZOE ROSE

Contents

Why Write a Book?	v
PART I	
Before You Write	1
The Fundamentals of the Book	3
The Outline	14
PART II	
The Writing	21
The Writing Plan	23
The Accountability Plan	28
Research	32
PART III	
The Manuscript	37
Your First Draft	39
The Read-Through	42
First Revision	45
Getting an Editor	48
Getting Beta Readers	53
Final Revisions	58
PART IV	
From Manuscript To Book	61
"Production Budget"	63
Proofreader	67
Formatting	72
The Cover	79

| The Back Cover | 86 |
| ISBN and Book Details | 96 |

PART V
It's Done! Now What? — 109

Publishing on KDP	113
eBook	118
Paperback	124
Audiobook	129
Hardback	134

PART VI
Launching Your Book — 139

Online Presence	141
ARCs	156
Newsletter	166
Ad Campaigns	169

PART VII
What Your Book Can Do For You — 173

Getting Publicity	175
Getting Business	183
Your Author Career	185
Afterword	187
Glossary	192
Resources	201

Why Write a Book?

If you've happened to pick up this book, there's an almost certain chance that you are looking to write one yourself. If so, congratulations! This is an excellent first step toward that goal.

If this is not the case, then I might recommend something else to read instead. You're probably not going to have too much interest in any of what we're going to be talking about here. I'm sure Stephen King has a new page-turner out.

This book is especially for people who want to write a book to build their personal brand. If that's you, keep reading. If you don't know what that is but want to know more, keep reading. If you want to write a nonfiction book but not for that purpose, still keep

Why Write a Book?

reading (but maybe skip Part 7 and the next paragraph).

If you're looking for the single best thing to build your brand, then you've made it. A book is the quickest and most effective way to establish yourself as a *published* expert in your industry, get attention for you and/or your business, and reap all the other benefits of being someone that wrote the book in your field. We'll get more into what exactly your book can do for you and how to get the most out of it later.

But everybody, and I mean *everybody*, has at least thought of writing a book at some point in their lives. Like skydiving or traveling to Japan, writing a book seems to be one of those generic sort of things that people have on their bucket lists. If it was easy, I'd argue that just about everybody would be an author.

And that's exactly why they aren't—writing a book is *hard*. It takes time, skill, drive, and desire that most people just don't have. A book comes from either a great story idea or someone with something worth sharing and, frankly, most people don't have that either. If they really tried to sit down and do it, the vast majority would not get past the first chapter.

I'm not being critical, by the way. I'm actually talking from experience. Avoiding any arrogance, I have some knowledge worth knowing within the IT industry from my many years of dealing with technical recruiters. If you don't know what they are, consider yourself lucky. If you do, then you know how those snakes literally get their paychecks from screwing over ours. They're a plague in our industry that cannot be avoided. You either have to accept that you will never earn what's fair or learn how to navigate them.

I learned how to deal with technical recruiters and get paid what I deserved and work to help others do the same. Based on my experiences of working both sides, I would even go so far as to say that I'm somewhat of an expert in negotiating with technical recruiters for a fair pay. It's been a long, long time since I've been paid less than what I know I'm worth.

The idea of people being paid less than they're worth drives me nuts. I've taught everyone that's worked for me, reached out, or just seemed like they needed to learn how to do this for themselves. People would tell me: "You should write a book!" And you know what? I thought so, too.

A book was the best way to get what I knew out to the largest audience possible. It'd let me help people I otherwise would never have been able to in fighting against the scumbag technical recruiters. If enough people working in IT knew how to get paid fairly, then maybe the whole industry would have to change for the better. An ambitious move, but I knew that a book was the best shot at what I wanted to do.

To be honest, I was pretty naive about the whole thing. Like most people, I really thought that writing a book was a matter of sitting in front of my laptop and watching the words fall right into place from the keyboard. I was running several different business ventures on top of having a family with no spare time in between, but that didn't phase me. If E.L. James could crank out a questionable best-seller on her Blackberry during her morning commute, how hard could it really be?

Turns out, *hard*—I sat on this project for two years and only finished about five pages worth of jumbled ideas. This was something I knew better than almost anyone. I've taught countless people this exact thing in person, over the phone, even through emails. And yet, when it came to writing it all down into a coherent chapter-by-chapter guide, I was stumped. How in the hell could that be?

Because writing, like public speaking or crocheting, is an acquired ability that takes time and effort to develop. We all think that we can write; we do it every day. But I'm not talking about everyday writing—that's far from what this is.

Why Write a Book?

There's a major difference between drafting an email to a coworker and *professional* writing, the kind that's meant to be read by many and which people can get actual value from. That kind of writing can only be done by those who have undertaken a considerable amount of time and energy to learn, practice, read, get critiqued, and revise until they are able to produce works worth any amount of someone's time. This is a specialized skill that some people base their whole careers on honing.

I'm someone that would honestly rather die than to accept a defeat of any kind. Seriously, I'm stubborn to a fault and have had to learn pretty much every lesson the hard way. I wasted two years before I was ready to admit that this was not something that I had the ability, skill, or time to do on my own.

That's when I went ahead and reached out to Zoe Rose, who ended up becoming my co-writer. Her work has been published in magazines, featured on a podcast, even produced by a theater company. Under a pseudonym, she's independently published multiple high-selling genre novels. I had the expertise of the subject, sure, but she had experience and a degree in the entire start-to-finish process.

We started working together on the project in June of 2020. By December, *Stop Getting Fu*ked by Technical*

Why Write a Book?

Recruiters: A Nerd's Guide to Negotiating Salary and Benefits was ranking in the top 20 business books on Amazon. I got to see my name listed just under geniuses like Chris Voss and (morals aside) Jordan Belfort, and that was pretty cool. Beyond that, I was able to do exactly what I set out to: share my knowledge with as many people who needed it as possible. And that's literally a dream come true for me.

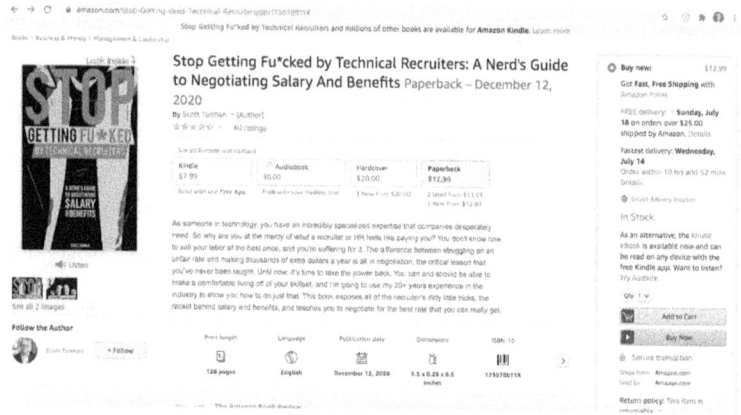

Our book on Amazon.

If what you know is worth sharing, you should write a book. If you have a unique experience or specialized skill set, you should write a book. If what you have or know could benefit others, you should write a book. If you're still reading this, then you probably have what you need to write a book. Write it.

Why Write a Book?

The best reason to write a nonfiction book is to share your knowledge or experience. After you've established that, what's stopping you?. If it's just something like a life goal, do it to prove to yourself that *you can*. However, if you're only looking to make money, then you're probably not going to succeed.

Failure is almost inevitable if profiting from book sales is your only goal. I've mentioned E.L. James and her smutbook being a runaway bestseller. She was a total amateur and a total anomaly. Her story is the fantasy that most people won't admit to having when it comes to writing a book. While the idea of making millions from the magic of your fingertips is incredibly appealing, the reality of writing a book for the first time couldn't be more different.

Out of those who set out to write, produce, publish, and market a book by themselves, very few are actually going to make any money directly through book sales. Out of those that somehow do, even fewer are going to make enough to cover the time and money spent to create the book in the first place. If you're looking at this endeavor as a way to make supposedly "easy" money, you are only going to be disappointed.

Besides, a book can do *so* much more for you. Having a book out in the world establishes your authority, builds credibility, and attracts attention. Like I've said, writing

Why Write a Book?

a book is hard, and not everyone can do it. That's exactly why it's so beneficial to those who can and do. The easiest way to make yourself an expert in your field is to simply get your name on a front cover.

A book serves to assist and promote your other endeavors. Backing up your expertise or skills with a book can rake in opportunities that otherwise wouldn't present themselves. A book can promote your product or service, bring in clients, even land you speaking engagements. Everyone thinks a book is impressive. A book will make *you* impressive and put you miles above your competition. You may not necessarily make any money from book sales but, trust me, you will make money from your book if you learn how to use it.

Don't get me wrong, sales are always nice to see. Everyone always wants to make money, sure. But the money I've made from my book has paled in comparison to everything else I've gotten from it. Once you see all that you can really do with a book, making money (from sales) is the very least of it.

This book will teach you how to write a book and what to do with it after. This means everything from the planning, first draft, editing, production details, publishing, marketing, and beyond. We have enough experience with writing and publishing books that we have worked out a complete process. If you are willing to put the time and energy into everything outlined, then you will be able to do this, too.

To get the most out of this endeavor, you must be willing to think of this as an investment and act accordingly. If you are willing to read this, you will learn how to write a book that pays off.

Before You Write

The Fundamentals of the Book

It may sound contrary, but you don't actually start writing a book by writing. There are several things that are prerequisite to even beginning the first draft. You'll only set yourself up for failure if you don't take the time to outline the book and create a writing plan.

Writing without an outline is like driving at night without headlights—you don't know where you're going and, eventually, you're going to end up with a disaster on your hands. To write something which people will understand and gain value from, it's imperative you plan out and organize the content into coherent, cohesive chapters that build on to your overall point of writing the book in the first place.

Before you write the outline, you have to establish some groundwork. This may seem a bit tedious, but you have to keep in mind that this is quite literally the foundation of your book. If you don't take the time and care to do this right, it's all going to fall apart; I can promise you that.

Before the outline, you need to address the fundamentals, the thesis, and a working title (or some name that you can attribute to your project). These three begin forming the basis of your book's plan. Planning out your book starts with defining your book's two fundamentals: the purpose and the audience. In our experience, this is typically the hardest part of planning out the book. Once you have these two things down, the rest of the prepwork tends to go pretty smoothly.

The first fundamental, the purpose, is exactly what it sounds like. What's your reason for writing the book? What do you want your readers to get from your book? The purpose needs to be specific and of concrete value. If you don't really know what your book is going to be about, neither will your readers.

My purpose for writing *Stop Getting Fu*ked by Technical Recruiters* is pretty obvious. I was tired of IT professionals getting ripped off and I wanted to teach them how to negotiate with technical recruiters to get the most that they can for their work. At the risk of sounding like an egomaniac, I'd say that's a pretty good purpose for a book.

My purpose is good because its value is self-evident and focused on one specific issue with the IT industry. If I tried to tackle *all* of the issues with the IT industry, the book would've sucked. Why? Having too broad a scope typically results in you talking about everything and never getting in-depth about *anything*. Everybody knows a little bit of everything, so there's no value in a book that doesn't dive into one specific thing, giving the reader insight they otherwise wouldn't have gained.

Defining the purpose often subsequently leads to defining the audience as well. If you know exactly what the book is about, then you usually know exactly who the book is for. Either way, it's critical to define your audience.

Like your purpose, your audience has to be *specific*. You need to understand that your book is not going to appeal to everyone. I'm sure (and take pride in the fact) that recruiters probably hate mine. Everyone doesn't need to read your book, either. If you did it right, your defined purpose is specific enough to disclude all but a very particular group of people that would find value in what you have to write.

Your audience, also, is not like a demographic. If you're familiar with some business-y terms, your book's audience is akin to a customer persona. But if you're not a business school graduate, there's no shame in not knowing what that means. Let me explain:

When running a business, a customer persona is the imagined, ideal customer for your product or service. It's a fully thought-out, made-up person that includes both who they are as a statistic and who they are as a *person*. This means not only their demographic information but also their values, motivations, fears—everything. You can create a customer persona to figure out who your audience is or who they should be.

Making a customer persona works because your book is ultimately, like anything else that's bought and sold on the market, a product. And as a product, you're just not going to be able to appeal to every person out there. Someone with dentures doesn't buy gum the same way that Suzy in Minnesota isn't going to read your book about maintaining a flower garden in the Southwest. If you try to get everyone, then you're really only going to miss who you actually need to target.

If you nail your customer persona, then you can understand who your intended audience is. That's who you need to write your book for. They are the only people that could actually derive value from your book, and so they are the only people that matter.

For example, the intended audience for *Stop Getting Fu*ked by Technical Recruiters* is booksmart-but-non-confrontational IT nerds who don't know how to negotiate. This audience tends to fall in the demographic of 20-something, middle-income people, but this is not the explicit audience. Not all 20-something, middle-income people are going to need this book. You need to consider the typical situation that someone would be in to need to read your book. They could all be from different ages, ethnicities, economic backgrounds, or other varied demographics, but are united by their common need for reading this book. That common need is what ultimately defines your audience.

Using your defined purpose, and keeping the audience in mind, you can now start developing your thesis. I'm sure you remember having to write a thesis in school. Here, it's similar but not exactly transitive. A thesis in a typical essay is "This is that because of a, b, and c." While that's a great starting point, it isn't enough to get your book started.

Here, you want your thesis to be more thorough. Remember, you're trying to write a book, not a two-page essay. The thesis for your book should define and discuss all of the points that you will present within the book and how they relate back to the purpose. This can be anywhere from one paragraph to one page in length.

To write your thesis, you can think of it as either the incredibly abbreviated summary of your book or as something that someone could use to understand what you're discussing, even at the most basic level. This can be challenging if you're an expert in your area, as it can be hard to simplify the depth of your knowledge. But by forcing what you know through a pinhole to sand down exactly *what* you want to write, you'll be able to then create a clear chapter-by-chapter outline of everything you want to get across—without getting lost in your own thoughts and ideas.

Writing the thesis is also a great litmus test as to whether or not you really do have a book to write. If you don't actually know what it is that you want to write, then you're not going to be able to narrow it down to a simple thesis.

A simple thesis is going to give you the thousand-foot view of what feels like a thousand-point subject. For example, let's say you have experience overcoming trouble socializing and want to share what you've learned. You may at first think that your book would be about how you had trouble making conversation, or not going out to social events after college, or maybe even back to how you never had a proper peer at home because your only sibling is a decade younger. These might all be aspects of the book, but it's not what the book is *about*.

So, let's back up a bit. All of those things point to one common problem: making friends. Through experience, you learned the solution. Here, the thesis is something along the lines of: how to make friends as an adult. The thesis is that main idea or one-sentence summary you can give about the book. It's that bare bones overview that you need to then be able to get into the meat of it.

I did say that you need *some* sort of title or thing that you can call your project before you get into the actual writing. A lot of people would disagree with that, and that's fair. Technically, you don't even need to have a title until you're more or less ready to publish. However, I recommend that you choose a working title or *something* that you can call it so that you can more clearly see and attach to the idea that this is your book, and you will be that much more likely to complete it. To quote Michael Wazowski, "Once you name it, you get attached to it!"

However, if you try to come up with a finalized title before you write, you may end up in a standstill of trying to pick the perfect title before you can write any of the actual content for the book. I'll tell you now that you likely won't settle on a good title until after you finish the book itself. If you wait until you have the perfect title to write, you are never going to finish the project. It's one of the ways our procrastination and perfectionism tricks us into not actually working, and you can't let yourself fall for it.

Instead, just treat anything you go with at this stage as only a "working title" to avoid that trap, even if that's the one you ultimately end up using. A good title will need to do these three things: *not* do any harm, be something marketable, and allude to the actual content.

What I mean by that first thing, not doing any harm, is really just that. If you pick an objectively awful title, you're going to make getting readers that much harder. A title that does harm is something that actively hurts your book and its potential.

It's hard to describe what a harmful title actually is. To best explain, we'll give some examples. Let's go back to that book about making new friends as an adult. That could be valuable to a lot of people, right? Here's how to destroy the book with a title and why:

- *How to Make Friends as an Adult by Being*

Genuinely Nice and Personable but Not Oversharing Personal Details. Aside from being ridiculously long, it more or less gives away the "secret sauce" of the book. Your book is essentially a solution to a reader's problem. If you tell them the answer on the front cover, then why would they even bother reading it?

- *My Incredible Journey to Becoming Sociable.* As obviously pretentious and silly as it looks, there's plenty of (horrid) books with titles like this. The biggest problem with a title like this is that it asserts the book to be grander and more epic than it could possibly deliver. Aside from that, an egotistical title is almost always inappropriate for our kind of nonfiction books. It's not a memoir, but the title would suggest otherwise. This just means that the people who should read it won't and the people that do read it are going to be disappointed and/or confused.
- *Friends.* It's overly broad, of course, but that's not even the biggest issue with it. Ever heard of a certain popular American sitcom from the 90s? Yeah, so has everyone else. And that's exactly what they're going to think of when they see this title. Now legally speaking, there's almost nothing stopping you from giving your

book the same name as other, more popular things. You can mistitle your book *Friends*, *Santa Claus*, even *The Bible*, and that's totally fine. However, the only people that will find your book are those looking for the sitcom, the Christmas legend, or their religious text. If they buy your book expecting one thing and get another, that's annoying at best. At worst, you're flirting with actual fraud.

As long as you steer clear of doing anything obviously bad, your title will likely not end up as some kind of non-example. But, to come up with a title that's actually good, you have to accomplish writing a title that both sells the book and gives an idea of what it's about without giving everything away.

This is a tall order, but I promise this isn't something to stress over. If you're looking for a shortcut to a good title, most good nonfiction titles tend to follow a sort of "formula" of a short but attention-grabbing title, followed by a longer subtitle that puts it into context. This way, you can have a marketable short title and then a subtitle that actually alludes to the content.

Back to that book about making friends as an adult: it took maybe a minute using that formula to come up with *Get Social: A Guide to Making Friends as an Adult*. While it could be better, it's not bad and it definitely does what it needs to. If you just spend some time playing around with this kind of structure, I'm sure you'll come up with something.

I feel obligated here to address the fact that I chose to include an expletive in the title of my first novel. Obviously, this has created some issues with mainstream publicity and has probably even prevented some sales. I've had random people contact me just to call me "a pottymouth" and criticize the provocative nature of my book.

But here's the thing—I fully understood that this would likely happen when I made the ultimate choice to go with that title and accepted any possible consequences that could come from it. My book was already going to piss some people off anyhow, so I may as well have really committed to it. You are always free to do what you want, of course, but you always need to consider all of the potential outcomes and consequences.

This all accomplished, you're ready to move on to writing the outline.

The Outline

I've already discussed the importance of all the pre-writing writing. Everything you've done so far is just so you can do the outline. I've said it before and I'll say it again: *you have to have an outline*. If you try to write your book without even a basic chapter outline, you are not going to finish the book. And if you somehow manage to, the book is going to suck—we can just about guarantee it.

Defining the purpose and the audience was so that you can write the thesis. Writing the thesis was so that you can write the outline, which will eventually turn into your book. Your thesis should detail out all of the points of your purpose. You use these points to organize into chapters. These points can be organized chronologically, by importance, relevance, gut feel, or however it makes sense for your specific book.

Some people like to take the approach that a nonfiction book is more or less an especially longform essay, and to structure it accordingly. This means presenting the central thesis in the introduction, making the major points of the thesis into chapters, and then affirming those points in the conclusion. But this is just one of many ways to do it, and it's not for everyone.

There's also what I call the "class" method. You've probably already seen it in a lot of nonfiction books. They structure their chapters into lessons that expand on one point to another in order to teach or convey the overall thesis. You can think of this outline structure as similar to a slideshow that someone would use to help teach a lesson or to make a presentation. If you're trying to specifically teach or otherwise explain something to your reader, this might be the best way to go.

Ultimately, your structure just needs to "flow" and your chapters need to "fit" together. These are pretty subjective terms that only you can really decide what they mean for you and your book. If you find that the chapters don't actually work in that order after finishing the first draft, then things can always be moved around. However, getting this part right during this stage will save you a lot of time later during the revision stage.

Most nonfiction books are going to follow this sort of outline:

- An opening chapter that introduces the purpose of the book, establishes your qualifications to write this book, and what the book will cover.
- Chapters that describe the points of the purpose, and build on the information presented.
- A conclusion that summarizes and reiterates your chapters. You may also use the conclusion for any relevant self-promotion.

Each chapter should more or less open with an introduction of what that chapter is about and then give a brief rundown of what the chapter will cover. Especially if your book's purpose and thesis involves subjects that the average person may not be familiar with, you need to work your way up to the more complex topics—you have to walk before you run. You might already know everything on your topic (obviously, you're writing a book on it), but make sure you cover the things which your reader might not be wholly familiar with before getting into the nitty-gritty.

A lot of people also tend to retain information and understand a concept better when there is some sort of narrative involved in the information presented. To this end, you may want to look over your chapter outlines and see if you have any anecdotes to include. Anecdotes, so long as they are relevant and contribute to what the book discusses, often work as great chapter openers, case examples, or segways into different topics. Readers respond to stories and if you choose yours right, anecdotes will greatly enhance the book.

A high quality anecdote will either demonstrate a key point of the book in real life, show the consequence of a problem presented within the book, depict a lesson in action, or other relevant experiences the content calls for. A main consideration is that it should relate to and reinforce what you are actively discussing within the book. You can't throw a story in only because it's funny, personal, or just something that you feel like sharing. If your chosen anecdote doesn't work to do *something* for the book, it's just going to be distracting and bad. If you feel compelled to share your life story, save that for a memoir.

And if you're trying to use your book to advertise yourself or your business, just don't be blatant about it.

Self-promotion is always understandable (at least in our book, pun intended), but it at least needs to be warranted and conducive to the book. It's the exact same way with the anecdotes. If it's not going to help with getting your point across, then it's just going to detract from your credibility and the reader's trust.

The best way to plug yourself in a way that can further your point is to include "how I/we do it" type of anecdotes. What do you or your company do that you can use as an example for what you're talking about in the book? That's ultimately what's most acceptable to include, and what might make a reader ultimately trust your service or product after reading.

I mean, hey, we're kind of doing the same exact thing here ourselves. This book is to teach people how to get their book done, but we also know that some people might realize that it's more time and work than they can really commit to. If they have the money, then they'll give us a call. You should adopt the same principle with your book—you're not trying to harass the reader to become a client, but you'll keep that line open.

There's no "right" number of chapters that you need to hit or adhere to to have a complete book. The number of chapters that you need in your book is exactly as many as it takes to thoroughly get your purpose and points across. Typically though, our observation has been that a nonfiction book will fall between five to ten chapters. Anything less or anything more is either done deliberately or because the author doesn't know what they're doing. So, if your outline falls under or exceeds that typical range, make sure that there is some kind of reason for it. If it's too short, you may be missing content. If it's too long, it's possible that you have *two* books in you. Just take that into consideration.

Similarly, there's no right word count to hit, but there does exist a *usual* number. From what we've noticed, that typical range for books like ours is somewhere around 20,000-40,000 words. You might need more or less to get everything across, but there's often going to be a lot of fluff to cut when a first draft goes past 40k.

For reference, a fiction novel is usually upwards of 50,000 words. Works of fiction need that length to establish characters, setting, conflict, complete narrative arcs, general prose/imagery/etcetera. We don't need to do any of that here. In fact, the less words that you can use to get your point across, the better. Your audience is reading to get what they need to know out of your book and that's it. Remember, you are asking your readers for both an investment of their money to buy the book *and* the time to read it. To attract the most potential readers and maximize the return on their time, try to make your book as short as possible without cutting any critical information.

But that's more of a note for the editing process. For now, you're ready to start writing the book.

The Writing

The Writing Plan

Ok, we might have lied—you don't actually get to start writing just yet. Up until now, you've been working to establish what your book is about and what it will cover. To speak in metaphors, you've drafted up blueprints to a really great house. But if you don't set a construction schedule, that house will never get built.

To speak in plain terms, you are incredibly unlikely to actually write your book if you do not set and stick to a specific schedule for writing it. You cannot "wait for inspiration to strike" or trust yourself to work on this intermittently. No matter if the words are flowing or you're suffering to make each sentence, writing is *work*. Luckily, the battle really only lies in planning and committing to a time to sit yourself down in front of the screen. Once you're there, productivity is almost always inevitable.

There are several different types of writing schedules to choose from. The best one for you personally is going to depend on your time commitment, availability, lifestyle, preference, etcetera. I'll describe a few schedules here, but you may even come up with your own entirely. The best schedule is really just going to be the one that works for *you*. So long as you stick to it, there is no wrong way to go about it.

The first typical writing schedule is to pick a specific time out of the day to write. This could be every day between 9:00am to 11:00am, 12:30pm to 1:00pm, 11:45pm to 12:00am, whatever is manageable and fits your routine. You decide that this is the set time out of your day that you will spend writing, and people tend to stick to it by working it specifically into their daily routine. Eventually, your brain may even be "cued" to be the most creative/productive during that time, but this is only based on some personal observations. This kind of writing schedule works best for people who need rigid routines, or who have a predictable day-to-day schedule and are generally sure that they will always be available during that time.

For those with varying schedules or otherwise need some flexibility with when they write, you may want to set a certain amount of hours per week to be spent writing. Say you decide to dedicate ten hours out of the week to your book. That could mean three hours one day, two the next, half an hour, then four and a half. As long as you stick to your weekly quota, you give yourself a lot of leeway with when you write and for how long.

My preferred schedule is to set a word count goal for each week or writing session. This way, you're setting goals for yourself that ensure actual progress on the book rather than focusing on the actual time spent in front of the screen. Word count goals force you to actually write rather than stare at a blank page during your writing time.

Word count goals also force you to break any procrastination habits. You might be sticking to your schedule of sitting down two hours a week to write, sure, but typing out and moving around two sentences during that time is not going to get you to the finished first draft. A lot of people fall into the perfectionist's trap of sitting and not writing until they can create one perfect paragraph to the next. Your first draft is probably not going to be perfect, but you have to get it all out to be able to fix it. That's the whole point.

All of these schedules are supposing that you can trust yourself to stick to it, though. If you're somebody that *needs* to have some pressure, whether real or self-imposed, your best bet may be to give yourself different deadlines to ensure that you hit every milestone of progress. If you give yourself until the end of every month to finish a chapter, you may not end up writing until then. But whether you're doing one paragraph a day or doing the whole thing the night before, you're still doing it, and that's ultimately all that matters.

Again, the whole point of setting a writing schedule is just so that *you actually do it*. That's literally all that's important here. You might only work for fifteen minutes a day or for one fifteen-hour stretch a month, it's your call, so long as you're actually writing and actively making progress. How you go about doing it doesn't matter in the least bit.

First-time writers tend to believe that you have to have a set place to write along with a time. In our experience, this is not necessarily true. While Maya Angelou rented empty hotel rooms to concentrate, Margaret Atwood writes wherever. Different people need different things, so whether or not you need somewhere specific to work or nowhere specific at all depends entirely on you.

Typically, the best place to work is obviously going to be somewhere where you can focus. What helps you focus varies. Some people need somewhere devoid of any distraction at all, while others need to have stimulus and background noise to keep their minds from taking a walk. That could mean standing at your kitchen counter, sitting at a busy coffee shop, in a Walmart bathroom stall, you name it. Wherever you can concentrate best is where you should go to do it.

One possible benefit of having a "set writing place" is that, like a set writing time, your brain may come to associate that place with writing, and going there might eventually work as a cue to your brain to shift into the mentality needed to be productive. Again, this is purely from observation, so you really just need to figure out what works best for you. We're all our own people, and we all have our unique "best" thing. Through trial and error, you'll eventually learn what yours is.

A writing schedule isn't the only plan you need to ensure that you make progress on the book. An accountability plan is just as important as a writing plan. We'll talk about how to create a situation to hold yourself accountable next.

The Accountability Plan

A plan for writing is a plan for success. But creating an accountability plan will very nearly *guarantee* that you actually get the thing done. Like any other major life goal, writing a book is something that a lot of people *want* to do, even *plan* to do, and still somehow fail to follow through with it.

Think about how most people's New Year's resolutions go. Someone says they want, for example, to lose weight. So, they pick a diet and exercise plan. Then, they buy workout clothes, fresh produce, sneakers, meal prep containers, everything. They pack kale salads for lunch. They even set their alarms an hour early each day to go for a jog before work. They thought this whole thing through. They've done everything they can to get it done and seem as determined as anybody could be to achieve their goal.

But by mid-January, most people are back to their old ways. Why? Because despite possibly thinking of everything else, they failed to keep themselves accountable to sticking to this daily progress. People that tell others about their goals or plans are a lot more likely to actually accomplish them because they have created a situation where they perceive a certain "social pressure" to perform. In this way, you basically have to trick yourself out of being able to quietly give up. The threat of a potential embarrassment is stressful, but effective. Believe it or not, a little stress is actually really good to keep up your motivation.

There are different ways that you can keep yourself accountable. Like a writing schedule, there is no one, single best way. The most surefire way to keep yourself accountable to completing the project is to share that you are doing it with others, but *how* publicly you go with this is entirely up to you.

Some people prefer to broadcast it to everyone by making social media posts, telling everyone they talk to, however helps. This works best if you need that public pressure to be productive, and a lot of it. This can also help you generate interest in your book, which may even later become your initial audience of readers.

However, if you'd prefer to not tell *everyone*, you might also consider picking one specific person to tell and to check in with on your progress. I recommend choosing someone that you hold in high regard or look up to, as you're likely to care more and be further motivated to not "disappoint" them. When choosing someone, always be sure that they are okay with taking on this kind of role. Some people may not appreciate having to assume that responsibility, while others may be excited to help out with your writing progress.

If you'd prefer to solely hold yourself accountable for completing the project and not telling anyone at all, then you may want to set up a private reward system for hitting certain milestones within the project. You could take yourself out for ice cream after each chapter, get that purse that you've been eyeing once you finish the first draft, go on a cruise when it's published, just about anything you want. Giving yourself permission to indulge in something may be what ultimately motivates you to get it done, and that's fine. Like I've said, whatever method works for you. Some people may actually even argue that not telling people about your project before it's done is better. Apparently, telling people can give you a "premature sense of achievement" before finishing it, and thus you'll be less likely to actually do it. Now I don't know so much about that, but it's something to consider.

Once you know what you're writing, how you'll organize it, and how you'll complete the writing, you are ready to actually write. You've done all the planning and *now* it's time to do it. Get to it, and then come back to this book once you've finished the first draft.

Good luck!

Research

If you've made it this far, it's likely you've committed to the writing process. You know what you want to write and how you're going to write it, and all of the information that you plan to present. But think back to your school days; before you could start writing, what was always the first step? *Research*.

Yes, you're an expert in your field, drawing on your years of experience. But even with your credentials, it's important to pull in other sources and support your points with research. It backs up what you're saying and shows that your book is more than a journal of your successes and can be of true value to your readers. Protect your brand and ensure a high-quality book by fact-checking your work using credible sources, practicing effective research habits, and incorporating a combination of your experiences and factual support.

If you present something as fact within your book and it's not a fact, it will reflect poorly on you, your field or company you are writing about, and worst of all, tarnish your brand. This can lead to loss of readership and likely some business. Writers who follow a traditional publishing route will often have a team of workers who double-check the information for you. But if you're following the independent or DIY route to publishing, it's up to *you* to hold yourself accountable. Again, it's ultimately your brand and credibility that gets tarnished if there's inaccurate information, so make sure you do your research!

In the age of the internet, the amount of information at your fingertips can be overwhelming. Practicing good research habits will help you find quality information more quickly.

Some basic tips include:

• Varying your search engine

• Simplifying your search term/phrase (less words usually equate to less erroneous results)

- Using quotation marks around a specific word or phrase that *must* be included (e.g., Catskills the location will also pull up results for cats and skills but "Catskills" will only collect results about the location Catskills in New York)

- Removing unhelpful words ('editing -video' will bring up editing results while excluding any mention of video editing)

- Steering towards more reputable sources (.org, .gov, .edu)

As you conduct your research, be sure to keep your notes organized in a document separate from your manuscript. That way, you can pull information without being bombarded by the content. This also proves especially helpful when you go to cite your sources—you won't waste time scrolling through your browser history if your notes include the URL to the article. Save yourself time later by being organized now.

When writing a book, you need to find the right balance between talking about your experiences and information from other sources. Some of this depends on your reason for writing the book. If you're writing to educate your readers on a niche subject and boost your business, then it's expected that the majority of the book will be about your experiences—that's okay. You have a snapshot of what life is like in your industry, and that knowledge is irreplaceable. We are merely cautioning you to support these claims with facts.

For example, I could make a general statement that independent publishing is on the rise in the United States, but if I claim that in 2011, independently published titles outnumbered the traditionally published market at nearly 2:1, I need to credit author, Justine Tal Goldberg, for this information from her article, "200 Million Americans Want to Publish Books, But Can They?" (*Publishing Perspectives*). These small efforts add immeasurable value and credibility to your book, which is especially important as a first-time author or as a relatively unknown expert in your field.

By this point, it's pretty clear how important research is for maintaining your credibility and protecting your brand's image. Be aware of setting yourself up in a trap, however. If you delay your writing or revisions for "research purposes," you could spend months or years continuing your research and making very little progress on your manuscript. Identify the key points you need to get more information on and then check them off as you go. This is where it helps if you've already written a draft and are simply doing research to polish and support what you've already written. If you're still in the writing process and you get stuck mid-section, simply insert "TK" into your paragraph (we'll explain why shortly) and keep the writing flow going. The worst thing you can do is to interrupt your progress every few sentences to double-check yourself —that's why we have a whole section covering revisions later. As for the letters "TK," there are very few words with these letters beside each other, making it easy to find the spot later using the command+f keyboard shortcut. This also works if you use a different key phrase you know to search for, test the waters and see what methods work best for you!

So, go forth and write about your expertise. The readers are showing up to hear from *you*, after all. Just be sure to give them the most accurate and pertinent information the internet can supply.

The Manuscript

Writing a manuscript does not actually equal a finished product. In this section, we'll be going over the entire revision process and everything it takes to turn your first draft into a real, well-written book.

Your First Draft

So you've made it this far. Congratulations! You've already accomplished more than most people ever will in writing a book. A first draft is a monster and a half to tackle, there's no getting around that. But you've hunkered down and pulled through, and that's downright impressive.

And sorry to break it to you, but the first draft is also only the *start* of the book.

The first draft that you've finished and the final book that you publish almost always turn out to be two very, very different things. Oftentimes the first draft and the final book are going to be *nothing* alike. The first draft is also often called a "rough draft" for exactly that—it's a rough iteration of your final product. The first draft needs some cutting down, building up, smoothing out, and a million big and little edits to get it right. The editing process can often end up being longer than the actual writing time itself, especially if this is your first book ever.

You've already put in a lot of time and effort to complete a first draft. As much as you might appreciate what you've got so far, I won't sugarcoat the likely truth of it: unless you're some one-in-a-billion brilliance (highly unlikely) that got it perfect on the first try (most definitely not), your first draft is probably not too good. So, if you want to make all of the time and effort spent worth it, you're going to need to put more time and effort into editing what you've done, that's just how it is.

Before you get to that, though, I highly recommend taking a break from the project entirely. Seriously, I urge you to take some time off and away from your first draft if you expect to do anything meaningful with editing it later.

You might think that's counter-intuitive. If you have all this work ahead of you, why put it off? If you've already got the writing momentum, why stop? That's exactly why—you've spent so much time locked into this project that you're probably suffering from a little bit of exhaustion and lock-in with what you've written so far. You're tired, you can't see past what you already have, and none of that is going to help you edit this book for the better. Taking time off after the first draft is vital to preventing burnout and will allow you to return to the project with "fresh" eyes.

Go ahead and put the draft away, take a walk, go to dinner, something that isn't writing the book—put some space between you and what you've accomplished so far. There's no right amount of time to take off from your book, either, but anywhere from one to two weeks is typical. Anything less than a week is not long enough, but anything past a month is going to take you out of the "space" to finish the project entirely.

Once you feel refreshed and ready to tackle the editing process, you start with a read-through. Let's talk about how to do that.

The Read-Through

A read-through is exactly what it sounds like. You read through your first draft in its entirety to get a holistic sense of where your draft is at, what you need to work on, and what's not working at all. It's hard to do this all by yourself, but a read-through is the start of understanding what needs to be done to get your draft to where it needs to be.

To get the most out of a read-through, I recommend doing it twice. The first time, you want to be reading for the content. Try to think from the perspective of a reader. They're obviously not going to know your subject as well as you do, and that's what they're going to try to get from your book. When reading your draft through that lens, does it seem like it completely presents all information necessary to learning the subject? If not, mark what's missing or where it feels like you need more. You'll figure out the specifics later.

On the flip side, where does it seem excessive or irrelevant? Is a definition of every single term really needed? Is that three-page story about your cat conducive to reinforcing a key point of the book? If not, cut it and see if you need to get something else there instead.

The second read-through should be for the prose, or the actual writing itself. I cannot overstate how important it is to read your entire draft aloud during this read-through. Reading the draft aloud will allow you to hear how the words work together and better judge where it doesn't "flow" right. Mark those sections. You'll fix it later.

Key word here is *mark*, not touch. You want to stay hands-off and avoid making any actual changes to the draft until you get to the first revision. You are in the read-through right now. Let yourself see what needs work before you actually jump in to doing it.

You should also be taking notes during each read-through. To stay organized, I like to use either a separate word document or even a physical notebook to jot down any and all thoughts that I have while reading.

Even seemingly miniscule ideas and impressions may prove incredibly helpful during your first revision, which we'll be getting into next.

First Revision

It's time to actually get your hands on the draft and make the first revisions. Let me start here by saying something that may be of some comfort:

Whatever you do here, it is very rare that you can make a first draft worse.

I mean, you have to really be trying to make it worse to actually accomplish that. While your first revision won't be the *only* revision, it will almost always be the start of seriously improving your draft.

How does revision work? Start by copying (that's keyboard shortcut command+c) the document of your first draft and then pasting (command+v) it into an entirely new document. That document should be labeled something like Revision_1 or something that makes it clear this is your first revision. Each iteration of your book should be in a separate, clearly-labeled document. *Never* keep working in the same document, as you want to save all of the earlier drafts. You want to always keep track of what you cut or change, because you never know if you might end up wanting to add it back in after all.

Now that you have the draft in a new document, you are free to go in and make the initial changes. The notes that you took during the read-through will obviously come in handy here, as they help guide what you need to edit.

Like the read-through, you should actually go through your draft several times and focus on one specific thing each time. First, you want to edit based on the notes and marked sections for the actual content of the book. After that, you need to go through and fix any issues with all of the "technical" parts of the book. That means the grammar, syntax, diction, and conventions —everything to do with the actual order and connection of the words within the book.

Finally, you need to edit based on the flow of the book. This edit is more of an intuitive, objective kind of thing. Basically, this editing is done to ensure that the book "sounds" right, and is done by again reading the draft aloud and marking or fixing where it feels off.

Of course, this doesn't mean that you're done. It's better than it was, sure, but you're not quite ready to take this to the printer just yet. You can't trust that you and you alone can make the book great. Honestly, without outside help and advice, it's unlikely that the book is going to be very good.

This is where we introduce your first outside resource in completing the book: the editor.

Getting an Editor

You probably already know what an editor is and the importance of getting one. Basically, if you can afford it, get one. This depends on your book's "production budget", which we'll be discussing in more detail later on.

Most people already understand that while a writer creates the content, an editor perfects it. They do this by first looking at the draft and commenting on any potential "big issues" that it may present. They will suggest *specific* edits, cutting certain content, expanding other parts, anything that will help further what's already on the page. An editor is incredibly valuable to improving a draft not only from their expertise and experience, but also from their ability to be objective with it.

They're not paid to spare your feelings or stroke your ego. What they say may not be what you want to hear, but they are going to tell you what you need to make it good. Sometimes that may even involve pointing out how a certain part you love just sucks. And that bites, no doubt. But a big part of writing a *good* book is accepting the feedback given and using it to improve; it's all a part of the process.

An editor is incredibly important to writing a good book. You should absolutely get an editor if you:

- Have the money to hire one.
- Want to ultimately publish a *good* book with clear writing and high-quality content. It is very hard to do this by yourself, especially your first time.
- Are open to suggestions and outside opinions.
- Are having trouble seeing the "big picture" of the book *or* need help fleshing out the micro-parts of the book.

So, basically, you should almost always get an editor to look at your book. However, there do exist a few very specific situations in which you may not necessarily need an editor. This could be a few things:

- You cannot afford to hire an editor. You also understand that not being able to invest in an editor will affect the overall quality of the book.
- You are a very stubborn person and/or are very fixed to the manuscript exactly how it is. You understand that not being open to outside help is going to affect the quality of the book.
- (Incredibly rare, probably impossible.) You have a full, complete understanding of how the book is supposed to be in its entirety and you have written it perfectly. There is no need for an editor because there is no room for improvement. Send your manuscript off to one of the Big Five publishing houses and collect your Pulitzer sometime next year.

The ultimate determining factor of whether or not you need an editor is your plan or intentions for your book. If you are writing this book for yourself and have no real goals for what this book will do, then you may not need an editor. But if you intend to get this book in front of others, then you definitely need to get an editor.

Where can you get an editor? It depends on what you are willing to spend on one. If you're not looking to drop any big bucks on this, then you may want to look at freelancing networks like Reedsy and filter by the price from low to high. Keep in mind that the lower-tier price will get you a lower-tier editor, but that doesn't mean that they will not produce good results. If you've got a little more to spend, you can find higher-tier freelancers or contact an established editing service.

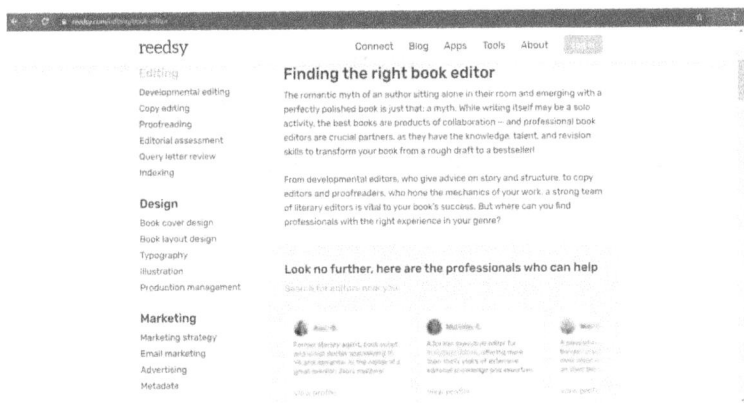

Finding an editor is easy on Reedsy.

Whichever way you go with this, always be sure to check their portfolio of past work prior to deciding on any one person. Their portfolio can tell you a couple of things. Most evidently will be whether they are actually worth the money, but you can also see whether they've done any work similar to your book before. If they haven't, I advise you to proceed a bit more cautiously.

The best editor for you is going to be someone that has already worked on books like yours, has a portfolio that indicates good work, and is easy to communicate with. That all said, an editor who fits this criteria can be anywhere from a few hundred to a couple thousand dollars. Like with anything, you get what you pay for here.

After your draft has been through a couple rounds of some tough editing, you're ready to take your book to your first audience: your beta readers.

Getting Beta Readers

The concept of a beta reader is probably not as universally familiar as an editor is. The benefits that a beta reader brings to a manuscript in progress is similar to that of an editor. A beta reader, however, is decidedly *not* an editor. They don't have any real expertise or qualifications like an editor, and that's usually why they're so beneficial for your book.

A beta reader basically acts as the test audience for your book. A beta reader will read your manuscript and provide feedback on it. Beta readers often provide some incredibly valuable insight, and they often do it for free.

Beta readers aren't meant to actively "fix" anything in the manuscript. All they do is give you their thoughts and overall impression. If you prepare some pointed questions for them, then they'll answer those as well. A beta reader is not a professional in the book business because your average reader isn't, either. Your book is a product and they are a great way to test the market before launch.

Always try to get beta readers within your book's audience. I recommend not having friends and family act as beta readers because they are less likely to give their most honest impression of the book. The best beta readers are going to be people who would actually read your book and who don't personally know you.

Luckily, it's not hard to find beta readers. There are loads of places on the Internet to look for them. Specialized Facebook groups and subreddits, CP Matchmaking, and Goodreads Beta Reader Group are just a few of them. Making a post on any one of these sites is bound to deliver at least a few interested readers, if not many (it should go without saying, but you want to get as many beta readers as possible).

CP Matchmaking is just one of many resources to find beta readers.

If you somehow don't find *any* beta readers or you need a very specific demographic, you can also pay to get some through freelancing sites. Beta readers customarily only take a free copy of the book in exchange for giving their opinion, and so offering to pay actual money will likely get you respondents if you so far haven't gotten any.

Beta readers typically used to get a free *physical* copy of the book. While that's still true for most professional publishing companies, people who read for indie authors usually only expect a digital version of the book.

The best way to make a digital copy that you can send to your beta readers is to create a PDF file of the book that you can then send out in an email list. For added security, you may want to consider making the PDF password-protected, only available for a limited period of time, or both.

You also need to understand that while a beta reader cannot legally steal your work due to intellectual property rights, there is a very slight chance that they may take or otherwise appropriate ideas/concepts/etcetera presented within the book for their own use. In these (again, unlikely) cases, there is not much that you can do. Since a paid beta reader typically has public ratings or makes income doing this, they are less likely to do this. Again, though, either one is unlikely to do this, but it is a very rare situation that you need to consider. If a beta reader seems fishy, it's best to leave them off of the list.

Now, what do you do with their feedback? You mostly just keep it in mind while doing your final revisions. My best rule of thumb regarding beta reader feedback is this: if the majority of them are saying something, listen. And if a suggestion sounds reasonable and/or makes sense, edit it in. If it sounds too far out or irrelevant, feel free to disregard. You have to ultimately decide whether your book is complete or not, so you ultimately have to go with your gut while revising with the feedback.

Oh, I did mention that you have to revise one more time, right? Luckily though, it really is the last part of the writing process.

Let's go ahead and finish this thing.

Final Revisions

Final revisions are for your editor's recommendations, your beta readers' feedback, and your last ideas. After this, you may take it to the editor for one last look or even to another group of beta readers for some final thoughts. But as the word implies, this is really the last round of edits and changes that you'll be making on your book. At this point, your book should be more or less done.

How do you decide when your book is definitely finished? How do you know when it's time to quit moving sentences around and actually put it out there? It's hard to say. Unlike a lot of other crafts, writing doesn't have any objective endpoints. There's no clear, definitive line to cross that takes it from a *manuscript* to a *book*. If you aren't careful, you could end up in a perfectionist's trap of constant small, meaningless edits and never actually complete the project. We've seen it happen plenty of times. When it gets to that point, it's never productive. Trust us on that.

Writing is an art. Art is subjective. If you want to create something *good*, then you have to accept that you will never write the perfect book. As Leonardo DaVinci said: "Art is never finished, only abandoned." There comes a point where you have to *decide* to be done with a book rather than definitively knowing that the book is done.

The only real way to decide if your book is done is by asking yourself this:

"Have I done the best job that I can to present a clear and complete message that will benefit my readers in some way?"

If you can honestly answer "yes", then it's ready. Or, at least, as ready as it will ever be.

So, congratulations! You've completed the manuscript. Next, we'll be talking about how to turn that into an actual book.

From Manuscript To Book

"Production Budget"

Much earlier in the book, we talked about goals and expectations for what your book will do. If you wanted to write a book simply to write a book, then you've already accomplished what you set out to do. You may be fully satisfied just knowing that you did it, and you can put that manuscript away in a drawer and move on with your life. And good for you.

If you were looking to *do* something with what you've written, though, then you're definitely not done. A finished manuscript does not equal a finished product. Completing the written portion is not the end of the road here.

Remember what I said about the first draft? It was all that time and effort, and it was only really the start of writing the book. Once you have the words written and ready, you're really not that far from being done. However, if you want to have a book out in the world that establishes your image as an author and actually works to build your brand, more time and effort is necessary.

Now, if you're strictly aiming to sell your manuscript to a traditional publisher, then you can skip right ahead to Part 5 (and truly, best of luck to you). In there, I'll get into why that's both:

a.) incredibly time-consuming and incredibly unlikely to pan out and

b.) almost always is never worth your time anyways

If that's your dream, then go for it. But if you have an open, entrepreneurial mind and want to get the absolute most out of your book, then keep reading.

Depending on what your goals are for the book, you may even need to put some money in addition to more time and effort into producing a complete book that is to your vision and full satisfaction. This is where the concept of a "production budget" comes in.

A production budget is the money that you have available and able to use for third-party services and professionals to complete the book. Your production budget is firstly determined by the money that you are willing and able to spend on this project, but you also need to consider what your goals for this book are. If you want to get this book in front of as many people as possible, then you need to make what I call the "packaging" (cover, interior text, etcetera) of the book as professional and attractive as your budget will allow.

While we're on book goals, I feel it is necessary to once again emphasize that in most situations, <u>book sales should **not** be the ultimate goal</u>. A first book is incredibly unlikely to generate profit directly through selling copies. And like I've said, a book can do so much more for you than that. If you're having trouble seeing what else a book could do aside from sales, just hang on, we'll get into that a little later on.

Your production budget is determined primarily by the money that you are able and willing to invest and your ultimate goals of the book. If you are looking to launch a business with your book, then you obviously will need to put in a bit more money than if you just wanted to generate a few more visitors on your blog. While it's not a matter of throwing money at your project and expecting results, a higher budget will enable you to ultimately *do* more, and it's important to keep that in mind.

The last big decisive factor of your production budget depends on your personal abilities and/or what production aspect that you want to focus on. For example, say you want to include some complicated graphics within your book. That's something that you would have to take to a high-tier layout professional, and so you'd need to put more money toward that. There may also exist aspects of your book's production that you actually can do yourself. If you're a graphic designer writing a book on graphic design, it's safe to say that you can probably manage doing your own cover.

There are three major production aspects that you will likely need the services of a professional to complete. These aspects are: proofreading, designing the cover, and formatting the interior text. I'll be explaining exactly what each aspect is and the different price points for them next.

Proofreader

An editor looks at a draft and tells you how to make it into a complete manuscript. Once you have that, a proofreader will make sure that the words on the page are, for all intents and purposes, perfect.

The most important thing that a proofreader will do is to make sure you don't look unprofessional. Beyond unprofessional, multiple errors throughout the book point to an outright amateur. It might seem silly, but those kinds of issues can and often will cost you readers.

Think about it. If your reading a book and se a bunch of typoss, spellin and gramatical errors, are you going to fully trust the authority and credibility of the author? Personally, I'm pretty much done with a book once I see more than one mixup of they're/their/there. Short of having an English degree, you can't trust your technical writing to be completely perfect and professional.

And even if you do happen to have an English degree, you are bound to have at least one typo somewhere in there. You have spent too much time in your manuscript and I can just about guarantee that you are not going to be able to see those micro-mistakes. If you want to be completely sure that your book is totally free of any technical issues, then you need to hire a fresh set of eyes to look through it.

Depending on your budget, you may also want to look into getting a copyeditor as well. It's probably obvious, but a *copy*editor is not the same as the editor that we talked about earlier. While an editor focuses on the macro vision of the book and will suggest big structural or foundational changes to the manuscript, a copyeditor is focused on the smaller details and specifics of the book.

A copyeditor will check for consistency, factual correctness, any potential legal liabilities, technical quality of the writing, and other formalities. A copyeditor's area of concern does somewhat overlap with an editor, but they ultimately are concerned with the smaller specifics rather than an editor's "big picture". A copyeditor is also similar to a proofreader in that they check the technical writing, but the proofreader solely focuses on this, and thus they are inevitably going to be more thorough.

Alright, that was admittedly a lot. Here's all that you really need to keep in mind:

- You always need to take your first draft to an editor.
- If you have the budget, I recommend that you take your manuscript to a copyeditor.
- Even if you don't get a copyeditor or an editor (again, highly don't recommend), <u>you need to get a proofreader</u>. Period.

A proofreader is essential. It doesn't mean that you're too incompetent to catch those small typos and errors. If you wrote your book, then your brain literally will not allow you to see those kinds of things. You have spent way too long looking at your book to be of any actual use. A proofreader ensures the overall professionalism of the text and is the last filter to remove those little signals that the book is "amateur".

You can get a proofreader at several different price points but again, like everything else, you get what you pay for. The cheapest (oftentimes even free) route is to run your manuscript through a program like Grammarly or some other automated service that detects spelling and grammar errors. This program has good reviews from its users, but it's mostly used for emails and other written communication. You have to understand that those kinds of programs aren't really geared for book projects and ultimately, it's up to you to trust a machine.

If you'd rather have an actual human person looking over your manuscript, then the most inexpensive option is likely going to be through a general freelance site like Upwork or Fiverr. While an individual freelancer is likely going to produce better results than a computer program, you have to keep in mind that the ones you will find on these general, low-price platforms are probably not going to be as good as a proven professional.

An experienced, qualified proofreader that does this professionally can be found through specialized search sites like Reedsy. These platforms are built specifically for professionals in the publishing industry, and so their work is often going to be the best that you can get. You're typically able to also see their past works, portfolio, etcetera, so you can make the most informed decision of who you ultimately go with.

At this point, we'll have to suppose that the text is pretty much perfect. Now let's get it looking just as good on the page.

Formatting

So, this is where we're going to get into some of what I call the "publishing minutiae", or the highly technical, somewhat-difficult-to-understand details of what it takes to put your words physically on the page. The first piece of the publishing minutiae is quite literally that—getting your text formatted professionally and neatly on each and every page of the book.

Putting attention into the formatting is integral to the overall quality of the book. If you trust that the text is going to upload and generate exactly how it is in your document, you will absolutely come to regret it. I promise. I can *always* tell when someone hasn't put any care into their text format, and I usually end up putting the book down. A poorly-formatted book is just as bad and as obviously unprofessional as a book with typos. Honestly, the look and layout of the words on the page can be just as important as the content itself. Sometimes, it can even be *more* important. I'm serious. You cannot skip this step.

There are two terms that are most used to describe this: typesetting and layout. Typesetting refers to and focuses on the actual arrangement of the text on each and every page to prepare for print. Typesetting is typically seen within big publishing houses. Layout, also sometimes called "mise en page" in fancy publishing speak, refers to the overall text formatting on each and every page. For an independently-produced book like yours, you're going to want to find a *layout* or interior text formatting professional.

You can get a professional either through a specialized site like Reedsy or through my reverse-search method: look at the text layout of independently-published books until you find one you like and then contact the author to find out who did the layout. And that's pretty much the two options, as a professional used by publishing houses are well out of reach for any independent project.

While I do ultimately recommend a professional, because this is obviously best done by someone with the experience and expertise to do it perfectly, I don't actually recommend outsourcing this for every book. If your production budget is on the lower end, then you're actually probably better off just doing some mediocre job that you can manage on your own. Let me explain:

Unless you have thousands of dollars at the ready, you are not going to get truly quality results. A *good* professional is going to cost upwards of four or five figures and nothing less. As they should—they have a highly specialized skill and should be paid accordingly. Moving text around might look easy the same way that, say, diving does. But like diving, there are a hundred little things that a professional has to learn and perfect to be able to deliver the results that we see. If it was easy, you could get a cheaper professional.

A lower-end "professional" listed on Fiverr or even Reedsy is more than likely just going to waste your money. I had to learn this the hard way for *Stop Getting Fu*ked by Technical Recruiters*. We contacted a supposed "professional" from Reedsy who had a good portfolio and seemed like they knew what they were doing. $300 and weeks later, we got absolutely disastrous results.

I don't even know how they managed it: chapters that started at the bottom of the page, random spacing everywhere, margins that changed by the page, even overlapping text. We would have been better off just using the original word document.

Of course, we didn't do that, but we weren't going to make another gamble. Instead, we did some research on how to do it ourselves. We learned that there *is* a DIY route here, but the choices aren't exactly abundant.

There are a couple of different programs to do the layout yourself. They all take time, intense concentration, and sometimes also money. If you have absolutely no money to spend on this (which, bad idea, but I understand), then you can see what you can do with Calibre or Reedsy.

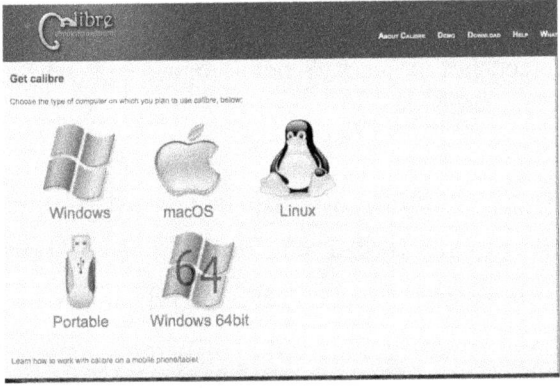

Screenshot of Calibre.

Calibre is a free, downloadable program that can convert a doc into a print-ready (PDF) and eBook (mobi) file, but it's not easy to navigate and the results are pretty dubious. Reedsy has a free program as well. It works through a mobile browser, so no downloading needed, and it's a little bit easier to use, but the results (again) aren't great.

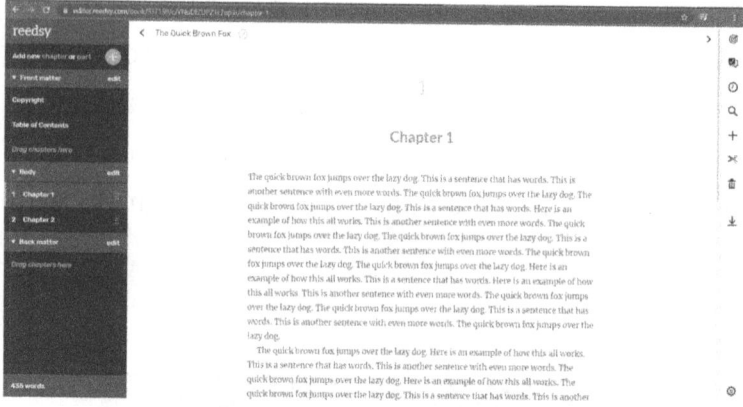

Screenshot of Reedsy's program, available online.

My ultimate recommendation is Vellum. Honestly, I'd go so far as to say that if you don't have the budget for a real professional, then you have to do it yourself. And if you have to do it yourself, you pretty much *have* to get Vellum. Don't think that this is an ad at all, by the way, because there are some considerable drawbacks to this program.

How to Build Your Brand with a Book

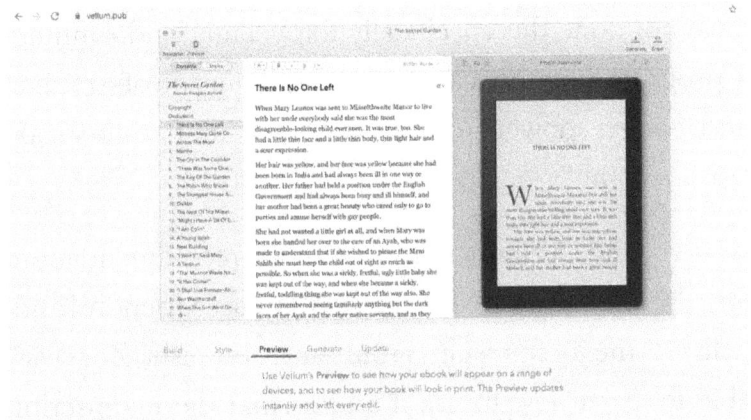

Screenshot of Vellum's website displaying a "preview" of their program.

First, it's $250 upfront. Depending on your production budget, that can be pretty steep. The upside to that price tag is that that's the absolute final price. You pay for the software license one time, and then you can use it for every project afterwards. If you're looking to write and publish more books, then you can see how it pays for itself pretty quickly.

Second, it's currently only compatible on a Mac. If you're not an Apple person, then that seemingly takes this out of your options entirely. However, like any other OS-exclusive software, there is a workaround. You can download a program on Windows to essentially run your device like a Mac. From there, you'll be able to download and use Vellum. Obligatory disclaimer: while I've done this for other programs, I have not done this specifically for Vellum, and running a program like this can take up quite a bit of storage and battery power, so take this with that so-to-speak "grain of salt".

As I've said, Vellum is my ultimate recommendation for the DIY route. It's incredibly user-friendly and delivers the best result. From start-to-finish, Zoe was able to learn how to use it and format the manuscript (both for print and digital) within a few hours.

Text formatting is essential. It's also often prohibitively expensive to get a decent professional and there aren't a lot of other DIY options. If you have to do it yourself, do it through Vellum. That's basically been our experience with it.

While you're working on the inside, you also need to put some work into the outside. Let's talk about the cover.

The Cover

I hate to break it to you, but it's true—people absolutely judge a book by its cover. You could have written something that's downright genius, but nobody is going to bother to even pick it up if the cover sucks. Inner beauty is all that really matters, sure, but people respond to what they most immediately see. That's why the cover is one of those key make-or-break factors. I've seen countless books that easily could've been bestsellers fail only because they had a crappy cover.

It's so rare for a book to succeed in spite of its awful cover that I can really only think of one example. *Interview with a Vampire* was published with such a wretched, cheap cover that it's an absolute wonder how it didn't end up with the generic faux-Gothic romance novels you see at the checkout lane of a grocery store. Somehow, though, it managed, and that is the one rare exception that I can point you to. And I can just about promise you that you are not going to produce the next *Interview with a Vampire*. Brad Pitt is not going to star in the adaptation if you just type your title in a default font over a stock photo (yeah, people seriously do that).

The actual first cover of the book in question.

Beyond the whole marketing and audience attraction aspect, your book simply deserves better than that. A good cover will reflect and reinforce all of the time and work that you've put into what's inside. Anything less is a disrespect to everything you've done so far and will basically negate all other efforts. If you want to make the book worth it, then you have to make that cover *amazing*.

What makes a cover amazing is ultimately a combination of the color palette, fonts, text, size, format, incorporated images, and all other visual elements that come together to create the overall design of the cover. The design should catch the reader's eye (in a good way) and allude to the contents of the book (subject, tone, all that jazz).

If you are specifically using the book to promote yourself or your business, then you may want to consider matching the color palette of your book to your website or to include your company logo somewhere on the design or even on the spine. This small visual detail will do a lot to tie yourself and/or your company to your book and all of the other aspects of your brand together.

You also have to remember that not only will you need a cover design, but you'll need that cover in the proper size and specifications for every format of the book. This means considering at least the eBook and paperback, but that could also include the hardback and audiobook. Every format has different dimensions and file requirements, and so you'll need your cover in multiple versions.

With all of that in mind, the benefits of a professional over some DIY approach to this is pretty self-evident. A properly-vetted professional cover designer will save you a truly incredible amount of time and actually be able to create something that you'll be proud to see your name on.

If you are not a graphic designer, then a part of your production budget *has* to go to getting a professionally-designed cover. If your production budget is an absolute zero, though, there exist a few (not great) options.

There are plenty of free platforms and programs available online that will let you make a cover. Amazon allows you to create a cover with their stock photos and basic templates while uploading your book. However, any and all of these options are almost certainly going to give you a cover that looks very much like the $0 DIY that it is. If you're going to try your luck with this route, then you need to keep that in mind and understand that the overall quality of your book is probably going to take a hit because of it.

The only hope here is if you have any proficiency in a pricey program like Photoshop. If you are willing to spend a considerable amount of time (as in, probably as much time as it took to write the book in the first place) to teach yourself how to use a professional program, then you *might* be able to make something decent.

But if you'd rather save yourself the time and stress and not ruin your book with a terrible cover, then you'll find getting a professional design to be more than a worthwhile investment. And it doesn't even have to necessarily be an *expensive* thing, either. There are plenty of lower-cost options that still produce high-quality results.

You can always start with freelance sites that advertise their rock-bottom rates like Fiverr. As I've already belabored in previous chapters, these freelancers are often going to deliver a "get what you paid for" type of product. However, I've seen some pretty good ones before and there's a chance you'll see some as well. My theory is that young professionals or graphic design students list themselves on there until they either realize that they can get paid more or don't find it worth their time. Either way, you might find something decent by gambling on a lower-tier freelancer.

Specialized freelance gig sites like Reedsy are often where the more talented freelancers and industry professionals go. They'll often have credentials and a portfolio to prove their high-quality work. Obviously they're going to charge substantially more, but it's almost always going to be worth that extra money. If you're looking for a very specific style, you can often find similar books produced independently through an Amazon search. Once you find one that you like and would want for yourself, you can usually find out who designed the cover and contact them from there.

Personally, my favorite way to get a cover is both reasonable (usually within the three-figure range) and lets me get something specific to what I want. 99designs is an online platform that connects you directly to a designer *or* lets you run your very own design contest. You specify the project, what you want, design preferences, etcetera, and get to choose from a range of different entrants. You can message the ones you like best with suggestions and feedback until you eventually get exactly the cover you want.

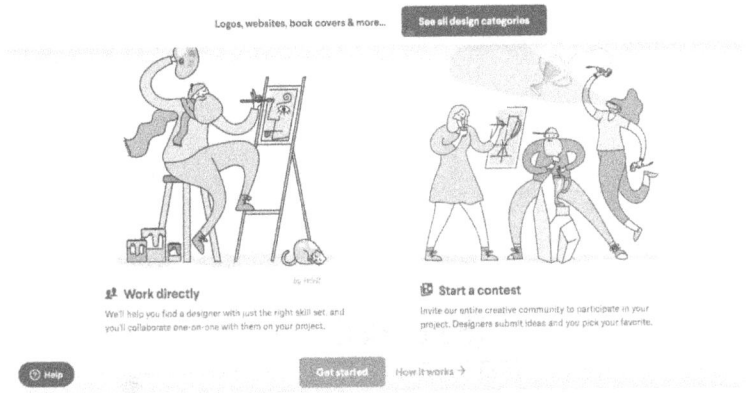

Screenshot of their website displaying the two options.

That may be what works best for me, but there's ultimately various ways that you can get a cover done professionally at various price points. If you're going to try to do it yourself, though, just remember that there's almost only one way you can do it—poorly. If you want to make all the work you've done worth it, then you almost always need to get someone else to do this part for you.

The Back Cover

A front cover is critical, of course. A front cover is what's going to get a potential reader's initial attention. But once they pick up the book or click on the title, what do they do next?

They're going to flip over the book or scroll down. Why? To read the back cover. The back cover is going to tell the potential reader what the book is about and what they will get out of it. Ultimately, the back cover is going to tell a reader whether or not they want to actually buy and read the book.

The back cover is basically the "sales pitch" of the book. You want to give the reader just enough to understand what the book covers, but you need to be explicit with what the book does for them. The brutal bottom line of your reader is this: *they do not care about your book*. All they really care about is what they can get out of it. For nonfiction, this means what they'll learn and how they can use it to improve their lives. This is why you can't just allude to what's in it for them. You need to tell them outright. What they will get from reading it is the entire selling point of the book. You have to make that loud and clear, or you're just not going to get readers.

A good back cover does a lot—gets the reader's attention, gives the essence of the book, matches the tone of the writing inside, tells the reader what they'll get out of it, and ultimately compels them to take the plunge. A good back cover also *doesn't* over explain and give the whole book away, (overtly) stroke the author's ego, promise more than it can deliver, or bore the reader.

This is admittedly a lot to accomplish in only a couple hundred words, but if you know your book (which, you should), then it's not really that difficult. Try to consider your book from the perspective of a potential reader. What's going to get their attention? What do you have that they most want? At the risk of arrogance, allow me to use the back cover of my own book as an example:

. . .

As someone in technology, you have an incredibly specialized expertise that companies desperately need. So, why are you at the mercy of whatever a recruiter or HR feels like paying you?

You don't know how to sell your labor at the best price, and you're suffering for it. The difference between struggling on an unfair rate and making thousands of extra dollars a year is all in negotiation, the critical lesson that you've never been taught.

Until now.

It's time to take the power back. You can and should be able to make a comfortable living off of your skillset, and I'm going to use my 20+ years experience in the industry to show you how to do just that. This book exposes all of the recruiter's dirty little tricks, the racket going on behind salary and benefits, and ultimately teaches you to negotiate for the best rate that you can really get.

Now, I know I can't avoid sounding like an egotistical jerk here when I say that this is a back cover that works. I mean, if it wasn't any good, then I probably wouldn't have gotten the results that I've had with it, right?

Here's why it works. The title already appeals to a very specific audience (IT professionals), and the first paragraph grabs their attention by describing the common problem within that audience (getting ripped off by technical recruiters). The next paragraph further describes this problem, and then the final paragraph explains what my book will teach them (how to negotiate) and how that will benefit them (getting a higher pay). At the same time, the diction and direct structure reinforces the sustained tone and style within the book, and my credentials are referenced (many years of experience in the industry). All of this works to appeal to the reader, their problem, and what they want. Ultimately, this is what gets the reader to make the buy.

There's no definitive formula to writing a great back cover, but there are some common markers. If you've got a hook, *just enough* of what the book's about, and exactly what the reader will get, then you're in a good place. If you want to make sure, you could always send out your back cover to your beta readers or someone who has never seen the book before and would have no reason to spare your feelings. Both groups will tell you if it's good and, if it's not, how to make it better.

Once you have a back cover that does the front cover justice, you can then move on to the last writing detail. Your readers are going to want to know not only about the book, but they're going to want to know a bit about *you*. You need to give them a concise, strategic biography to tell them exactly what they need to know.

Your author bio isn't just a chance to get to unabashedly talk about yourself (which most people love to do), but it's also where you get to establish your credibility for writing the book in the first place, and legitimize everything you've written. An author bio is surprisingly important. Not everyone looks at it, sure, but those that do often use it as the deciding factor for whether they'll actually read your book or trust what you've written. A bio that doesn't affirm your authority can do a lot more harm than you think.

So, a good author bio establishes credibility, but it's also *concise* while doing it. The personal information that you include should only be chosen if it works to build your authority on the subject of your book, and that's it. I'm sorry, but nobody really cares about your band in high school or stamp-collecting hobby. Anything beyond mentioning your family or where you live is only to serve your ego, and people are going to clearly see that.

But a good author bio is not a summary of your achievements, either. You might have won the spelling bee in third grade, but what does saying so do to legitimize your book on the responsible ownership of pet snakes? Unless it's something that you're truly famous for and could draw in readers, any irrelevant accomplishment should be left out of your bio.

If you're still unsure of what to write, you can always go check the author biographies in the back of a book similar to yours to get an idea. Also, what do you notice almost always accompanies the bio? A picture of the author themselves. You'll need that, too.

A picture is worth a thousand words and an author photo helps build your credibility much more than you would ever expect. Humans are highly visual creatures. People are going to respond to what they can see, and a picture of you and your shining smile will earn you way more trust than you'd even believe.

You might be shy, insecure, or what have you about your image and think showing yourself isn't necessary or worth it. You might even think I'm a hypocrite. I'll be the first to point out that I did not include an author's photo in my first book, and I'll be the first to admit that that's one of the biggest things I would change if I was to do my book over.

But hear me out—the point of your book is to get people to listen to what you have to say and, past that, to connect with you and your brand. It's not about being good-looking or attractive, either. It's about seeing *you*, and there's nothing like getting to see the author themselves to get the reader to connect to them.

But it can't just be any photo you have. A duck-lipped selfie is going to do a lot more harm than no picture at all. A good author photo is professional, well-lit, and stages the right "tone" to accompany the voice you present within the book.

Now, what exactly does that mean? It depends on the book. For example, if your book is a humorous take on parenting, then it'd probably make sense to wear something clean-casual and maybe even include your child (or, if you've decided to write a book about parenting without being a parent, finding *a* child will probably do). If you've written a slick business book, wearing something sharp, going for a black-and-white color grading, and not smiling *too* big would help evoke that serious, authoritarian image you need. Like I said, what you need to do varies from book to book, and so you really just have to consider what it is that you've written, what image you need to project, and your brand.

But there are some basic "don'ts" in taking an author's photo that apply to everyone:

- Don't wear anything shabby, overly bright/patterned/otherwise distracting, or ill-

fitting. You want what you wear to be put together but otherwise inconsequential.
- Don't take the photo from any odd angles. Keep the photo from the waist or neck up and close enough to see your face without seeing *everything* (pores, Crow's feet, etcetera).
- Don't have a distracting background. Have a background that fits with your "tone" (which could mean on a street, by a building, etcetera), but don't make it all that anyone can look at. If there are people in the background, make sure that they can't be individually identified.
- Don't use any filters. This should be implicit, but people want to see an authentic image of you. Filters and digital alterations (beyond fixing red eyes and blurring the errant pimple) are obvious and will absolutely lose you some trust with your readers. Again, the point of an author's photo is not to sell copies by showing off your looks, attractive or not. It's about showing yourself to the reader to build some trust and get them to connect to *you*, not some unreal image that you put in place.

Like I said, I've learned from my first book. As you have already probably noticed, we both include author photos for this book. If you haven't, here is mine again.

My author photo.

As I've recommended for most everything else, a professional is best to get this done right. That's what we both did. But if a photographer is completely out of your production budget, you can take a decent picture on your own so long as you avoid the "don'ts" and follow some of the basics of photo composition (lighting, positioning/angling the camera, maybe even playing around with color saturation or other adjustments).

However, if you can get a professional, then absolutely do it. You can find plenty of local photographers through a Google or Facebook search. Look specifically for "headshots" or "professional photos" and check that they have similar shots in their portfolio. Typically, they shouldn't run you more than $100 at the most for a simple headshot.

You can sometimes even find a photographer just starting out who would be willing to take your picture solely in exchange for having something to put in their portfolio, although you have to understand that they may not produce a photo quite as good as a paid professional. Free photographers are more likely to be found on certain Facebook groups, Craigslist, and even on model listing sites. Be wary as well that they are "legit" and fully understand what you are trying to get, as scams and scary situations can happen. So long as you're safe and smart about it, you'll be fine.

With the author photo done, you have just about every detail of the book ready to go. Just a few more little technical details to go.

ISBN and Book Details

If you've ever flipped to the back of a book, you might have noticed a barcode and some random numbers and letters printed on the bottom. That thirteen-digit code is the International Standard Book Number, or the ISBN. An ISBN allows someone to look up and see all of the details of your book, connects your book to multiple databases, and increases the overall visibility of your book. Basically, if you want your book to really cement your book's legitimacy, then you need to get an ISBN.

An example of an ISBN and barcode.

An ISBN is one of the biggest factors between an "amateur" or clearly self-published book, and a "professional" or respectively independently-published book. If your book does not have an ISBN, then the distribution platform that you upload your book to will simply assign a barcode and code used only for its in-network identification. Amazon will assign what they call an ASIN, or an Amazon Standard Identification Number. These codes are incredibly obvious and are probably the biggest giveaway that a book was not published professionally. An actual ISBN will make it nearly impossible for the average reader to know whether or not you published your book independently (nor, considering all of the other professional aspects of your book, will they care). An ISBN is what takes your book from a personal project to a legitimate product that can be identified, accessed, and ordered based on its internationally-recognized identifier.

```
Product details
   ASIN : B081H2GM48
   Publisher : Best Seller Publishing, LLC (November 15, 2019)
   Publication date : November 15, 2019
   Language : English
   File size : 9484 KB
   Text-to-Speech : Enabled
   Enhanced typesetting : Enabled
   X-Ray : Not Enabled
   Word Wise : Enabled
   Print length : 134 pages
   Lending : Not Enabled
   Best Sellers Rank: #199,154 in Kindle Store (See Top 100 in Kindle Store)
      #14 in Outsourcing (Kindle Store)
      #29 in Outsourcing (Books)
      #560 in Entrepreneurship (Kindle Store)
   Customer Reviews: ★★★★☆    26 ratings
```

Notice how this book's product details lists an ASIN instead of an ISBN. This is the so-called "tell" of a book that was not professionally published.

As I've said, an ISBN will connect your book to multiple databases. This increases the likelihood of your book being found by others, which in turn may increase sales. And if you have any dreams of getting your book into a brick-and-mortar store, you *need* to get an ISBN. No bookstore will carry a book without one.

Even if you're looking to stay strictly online, an ISBN is a huge benefit to your book. Trust me—just get one.

The easiest way to get an ISBN is through Bowker Identifier Services. Their website will walk you through the ISBN-assignment process and allows you to register either as an individual, publishing organization, LLC, and everything in between.

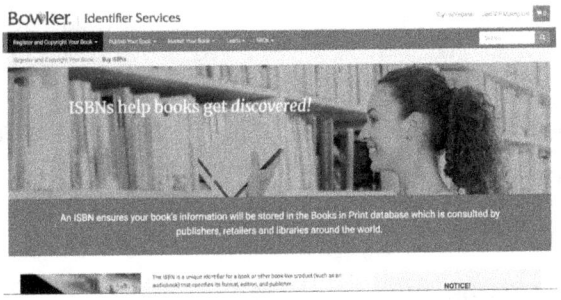

Screenshot of Bowker's website.

To get an ISBN, you'll have to fill out some details about your book. This includes the author credit, description, publication date, etcetera, which are all things that you should already have established. But there are three things that you may not have gotten figured out quite yet: the keywords, categories, and book price.

To ultimately determine the best categories to place your book in, I recommend by first starting with the best keywords for your book. Knowing the multiple different words that you can attach to your book can be helpful when choosing the two or three "big" labels and understanding your book's general placement.

Keywords for your book are a lot like Search Engine Optimization (SEO) keywords—we'll go more into how that works later on. You want to choose the words and phrases that people are looking up and what can get people to your book, of course. But like anything SEO related, the best keywords are going to be niche and very specific to *your* book. If you've written a recipe book for cat treats, "cats" and "recipes" are a good start, but "cat treat recipes" will be what really connects your book to your readers.

And while attaching a bunch of frequently-searched words and phrases might get you some attention, it's unlikely to get you any readers if those keywords aren't directly related to your book. If anything, you'll lose out on those who actually want to find your book. It's a quality versus quantity thing here, with the added bonus of even possibly getting flagged for spam by abusing keywords. So again, choose what really fits with your book. There's no other way to "win" more readers than those who actually need and want to find the information that you're giving out here, anyways.

It can be challenging to pinpoint the few allotted keywords that both fit your book *and* are most strategic for SEO. The best way to do this is to start by sitting down and listing out each and every word that fits your book. I recommend doing this with an actual pen and paper because it can get messy, but typing it out is fine too.

You want to build up a decent-sized list of words so that you can then narrow it *down* so that you'll eventually end up with only the best terms and phrases to fit your book. The first part of this culling process is to simply cross out or delete the terms that are the least related to your book. If you did the brainstorming part right, you'll definitely have at least a few useless ones on there.

After that, run the keywords left through any one of the keyword research tools available online to see their individual search frequencies. I personally use the Keywords Explorer available on Ahrefs but again, any one of the many online tools will be able to tell you how often a word or phrase is searched up. Write or type those numbers down next to each word. From there, pick the terms with the highest numbers *et voila* —you've got your keywords.

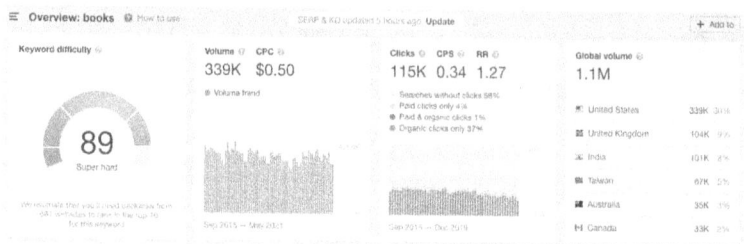

The Keyword Explorer Tool on Ahrefs can tell you how often a certain word is looked up, among many other things.

And by knowing the micro (the keywords), you'll be able to figure out your macro (categories). Not to stress you out, but putting your book in the right categories is absolutely critical to reaching your readers. Again, no pressure, but it's one of those huge make-or-break factors that you can't really just gloss over.

But it's really not something to agonize over. Choosing the right categories is just a matter of understanding what your book is and doing a little research. If you've managed to get as far as writing the book, then you definitely already know what your book is and who it's for. Out of all of the categories and subcategories available, you probably see a couple places where your book would fit.

Like keywords, the best categories are ultimately going to be strategically chosen but still *specific* to your book and its contents. Your book on making cat treats might fall under "general cookbooks" or "pets", but that's not necessarily the best place to ultimately put it. There's a sweet spot that you want to hit between where your readers can find you and where your book can rank. Let me explain.

Amazon ranks books based on sales and reviews both for the entire store and for each category. Amazon's algorithm picks up and pushes books as they're ranked. So the higher that your book ranks, the more visible it becomes to potential buyers. When a book hits the top spot in either the store or its category, they get that coveted "best-seller" banner. Getting that banner elevates your book's prominence and also gives you the bragging rights of being a "best-selling author".

This book picked up enough traction to get an "Amazon Charts" banner.

Amazon's "#1 Best Seller" banner.

Some authors will even add a banner or badge on their book cover to advertise their book's "best seller" status.

However, a best seller isn't always really best-selling. Some independent authors have learned to game the system by setting their book in the most niche, least competitive category in order to get that banner. There even exist marketing agencies that charge a truly ridiculous sum just to do this.

And while yes, you can easily do this to technically become a "best-selling author", this tactic will end up killing your potential for readers. It's similar to the keywords. By setting your book in the most niche category possible, those that would actually want to read your book won't be able to find it. Unless you'd rather get that imagined prestige over a real readership, it's best to put your book where people would actually be looking.

To that end, you need to think like a reader. Where would you look for a book similar to yours? Click around and write down all of those potential categories. From there, it's more or less the same process as the keywords.

Like keywords, your categories should ultimately strike a balance between best fitting your book *and* best calculated for ranking as high as possible. What this means here is that you need to find categories with less competition but still somewhere that a reader would actually be searching in. By thinking like a reader, everything on your list should satisfy being an actual "fit" for your book. Now, you just have to figure out where it'd be easiest to rank.

To start, take all of the categories you have listed out and find the #1 book in each one. In the book's product details, you'll be able to see their "best seller rank". The first number is that book's rank in the entire Amazon store for all books. Write that down.

> **Product details**
> Publisher : BrightRay Publishing (December 12, 2020)
> Language : English
> Paperback : 128 pages
> ISBN-10 : 173578611X
> ISBN-13 : 978-1735786117
> Item Weight : 7.8 ounces
> Dimensions : 5.5 x 0.29 x 8.5 inches
> Best Sellers Rank: #487,997 in Books (See Top 100 in Books)
> #475 in Business Negotiating (Books)
> #6,983 in Personal Finance (Books)
> Customer Reviews: ★★★★★ ~ 41 ratings

The "Product Details" section as it looks on
an Amazon page.

Once you have that number for each category, you'll then want to use a "best seller calculator" available from a simple Internet search. There's one for Kindle and one for books, so make sure that the calculator and numbers that you're using correlate with the market that you want to rank in (though my recommendation is to always go after the Kindle market, as it tends to be easier to get sales there anyhow).

The calculators available online will tell you approximately how many books any particular title needs to sell each day in order to maintain that overall rank in "books". The lower their calculated daily sales are, the less competitive their category is. These low-competition categories that fit your book are what you want to choose.

The last big detail that you have to figure out while registering for your ISBN is the price for each format. Since sales aren't the goal, your knee-jerk reaction might be to go low. We're going to talk about exactly why you shouldn't make your book dirt-cheap in a little bit, but here we're going to talk about why you should actually list it for *higher* than you should.

Figuring out the best price is as easy as looking at the common price point for books of similar length in your categories. But when filling it in on your ISBN, you may want to consider bumping that number up about $5 to $10 more than you plan to really sell it for. It seems incredibly counter-intuitive, but it's actually just a common business tactic.

By listing it higher on the ISBN and pricing it lower on Amazon, people often get the impression that they're getting "a good deal" on Amazon. And when people think they're saving money, they're more likely to actually spend it. Amazon will often even highlight this fact by marking the price as "reduced" so all of your readers can see it. Basically, it's creating a fake sale to make *real* sales, if that makes any sense.

You can buy either one ISBN at a time or multiple for a reduced rate, which I recommend if you are planning on publishing multiple books. You also need to keep in mind that you'll need an ISBN for *each* format that you publish your book in, so that could mean that you'll be buying anywhere from one to four ISBNs.

I won't sugarcoat it—ISBNs are a little expensive. Bowker's deal is ten for about $300. Depending on your production budget, that might be a little steep. It's certainly nothing to blink at. But once you have it, it's yours and yours forever. An ISBN is that last detail to solidify your book as a truly professional product, and that is well worth it.

It's Done! Now What?

If you've come here after skipping the previous section, I get you. You want to see your book traditionally published, see your name on a shelf in a store, all of that. And while I respect that classic writer's dream, I unfortunately have a harsh truth for you:

Traditional publishing sucks.

And it has for a long, long, long time. There was once a time where print was the only media out there and writers could make an honest living off of their royalties alone, but that time's been over since at least the age of the Internet. And let's be real here, most people don't choose to read in their downtime anymore. The market is *tough*, to say the least, and publishers know this.

Traditional publishing houses take no chances on new authors. They need a guarantee that you're going to make them money. If you're not a celebrity or have a manuscript that just happens to hit the current news cycle on the head, there's essentially a zero chance they'll even consider you. If you're still stuck on it, though, I'll walk you through what you need to do to get traditionally published.

First, you'll have to start by writing a query letter and sending it out to various literary agents. You have *literally* no chance without an agent. You'll get a few replies to read the first few chapters, the full book, maybe an offer. Usually, though, you wait and wait for some very polite rejections.

But say you get an agent. Cool. And that agent pitches you to a publisher. The publisher likes you. Very cool. They force you to make a bunch of changes that you don't want, go with covers and designs that you might even *hate*, and then get you to do a majority of the marketing work. You might get a four-figure advance, *maybe*. And if your book makes back that advance, then you'll start getting royalties. Usually, those look like less than a quarter of the sales. Sometimes that's even less than 10%, I'm not kidding. And again, that's supposing that your book even makes back the advance. They most often don't.

A publisher will take care of all of the publishing details and will give you money upfront for your book. But a publisher will also take away all of your creative control and push all of the advertising responsibilities onto you in exchange for a ridiculously small percentage of the sales. So unless you're Stephen King and can pump out multiple best-sellers, you're not making anywhere near the time and work you've put into it.

Again, this is all supposing that you can even get that far. Most people struggle for years just to get an agent. Some never do at all. For the average person, the chances of getting traditionally published and actually making money out of it are about as good as winning the lottery.

There's a couple things that might push the odds more in your favor: if you're well-known enough to the general public, famous within a niche audience, have a large social media following, have a one-in-a-million stellar manuscript (and I'm sorry to say, but this is unlikely), or have managed to finish a manuscript just in time for a related topic to get big in the news. If any of that applies to you, then traditional publishing may work out better for your particular manuscript. At this point, there isn't anything left for you in this book. Goodbye and good luck, I guess.

If, however, you're the average person who wants to actually see their book out there and get the most out of what they've done, then read on.

Publishing on KDP

First of all, I want to get this out of the way: it's not self-publishing. I hate that term. That's a word exclusively reserved for vanity and ego projects. You have written something of value, and you have opted for *independent publishing*. Huge difference.

Hell, I would go so far as to say that if you're willing to struggle and wait for *years* to get little more than your name on a "traditionally published" book, then that's the real vanity project. An independently published book is an entrepreneurial endeavor.

Independent publishing has always been around. It has its roots in a response against the system, from Martin Luther to punk zines. It doesn't mean you couldn't do "real" publishing. There have been plenty of talented people with good works who couldn't find a traditional publisher (or didn't like their crap offer) who ended up having a breakout success with independent publishing.

For the entrepreneurs and for those that want to do it their way, independent publishing is the best solution to all of the problems with traditional publishing. And with the increasing popularity of eBooks and Print-on-Demand (POD) distributors, it's easier than ever to get your book out there and into the hands of as many readers as possible.

While traditional publishing houses will take care of and pay for all of the publishing minutiae, going independent means that you retain complete creative control of your book from cover to cover. You might have to go find a designer and pay for the cover yourself, sure, but you're getting *exactly* what you want. You have the final say on your book and nobody else. For a lot of people, that's more than worth the upfront out-of-pocket.

You don't get an advance with independent publishing either, of course, but then you also have no advance to have to make back. You start getting royalties with the very first sale. Speaking of royalties, you usually get anywhere from 30-70%. That's significantly more than any traditional publisher would ever offer.

Nowadays, a traditional publisher pushes all of the advertising duties onto the author without any extra advance or royalties. With going independent, you at least get to see a direct payoff with that work.

While there exists many different distribution platforms available for independent publishing, there's only one to really consider. Hands down, Amazon's Kindle Direct Publishing (KDP) is the best place to publish your book. For better or worse, Amazon has come to dominate just about every market and can reach more readers than all of their competitors combined. If you want to put your book on other platforms as well, go right ahead, but KDP is absolutely *the* platform.

KDP's login page.

If you want to publish your book through KDP and KDP only, then you have the option to enroll your book in something called "KDP Unlimited". This is a program that allows readers to access any and all books that have enrolled into KDP Unlimited for a monthly subscription fee rather than paying for each book individually. It's a great deal for someone that reads a lot, and it can also be a good deal for you as well.

Notice that the Kindle edition has a price listed at $0.00. That is because this particular book is enrolled in the Kindle Unlimited program.

Authors get a payout depending on the monthly "pot" of what the subscribers have paid and your Kindle Edition Normalized Page Read, otherwise known as your KENP. This is basically the quantitative calculation of how many total pages were read of any and all of your Kindle Unlimited books by Kindle Unlimited subscribers.

A lot of authors have argued that essentially paying you per page read is unfair to those who have enrolled their graphic novels, children's books, and other works with a typically lower overall page count. However, Kindle Unlimited is an extra option to possibly generate more readers and more money that I recommend for those who either aren't looking to publish their books on other platforms anyhow, have longer books, or who simply want to reach as many readers as possible.

If you are looking to get as many readers (and sales) as possible, then you need to consider having all formats available. At the very least, you need to have an eBook and a paperback available. Let's talk about why.

eBook

I'm sure that you're already aware of the prevalence of eBooks. Like most other things, a lot of people now choose to read exclusively in the digital format. And that totally makes sense, too. The convenience of having all of your books on one lightweight device and the ease of accessibility is inherently appealing. If you had to choose to publish your book in only one format, the digital version is probably going to appeal to and reach the largest portion of your possible audience.

Some authors have reservations about creating a digital version of their book. This is all based out of ego and the idea of legitimacy. Their thinking is something like this: "Well if it's not in ink and paper, then it's not a real book. And if it's not a real book, then I'm not a real author, so I won't even bother with that."

You may find this line of reasoning rightfully erroneous, but you might be surprised to know that there's actually quite a few people who believe this. One of the biggest publishing mistakes that a first-time author could make is to reject the potential of a digital reading audience because of their own vain idea of what a "real book" is. By being strictly in paper print, they're missing out on a majority of their potential readers.

And anyways, an eBook *is* a "real book". All of the most famous and widest-read books—*Harry Potter*, *To Kill a Mockingbird*, *War and Peace*, *The South Beach Diet*, everything—are available both in print and digital forms. All professionally-published books are available in both. In fact, having both major versions available will only contribute to establishing your book as "legitimate". So if you do have any of those notions of what constitutes a "real" book, you need to understand that an eBook makes it *more* real.

Aside from that, releasing your book first in its digital form can be incredibly advantageous to getting your title ranked in its categories. eBook rankings are typically a lot easier to get compared to paperback, and so it can greatly increase your chances of getting that "Best-Seller" banner over your book's Amazon page. But we'll talk more about how to get rankings later on.

Setting up and publishing your eBook through Amazon's Kindle Direct Program is as simple as filling out information that they walk you through and then uploading your cover and text. And if you've set up your book's ISBN, then you've already done all the work for setting up your book.

What the setup page looks like.

Most of this information can be copied and pasted from what you had to fill in while registering your ISBN, including keywords and categories. Remember to also double check that you're providing the right ISBN number for your eBook, as each ISBN is specific to each book format.

You should also have made sure to get your cover correctly rendered for all formats. The biggest difference between your eBook and paperback cover is that your eBook won't have that back cover and all of the information there. That's not to say that your digital readers won't see it, though. Instead, the book description and author bio will be available directly on the sales page.

Another big difference between your eBook and your paperback version is that an eBook is generally going to be a lower price. You've already done the research for the best price point of your book, and I again recommend that you set your price on Amazon for lower than what's listed on the ISBN to give that impression of a "good deal".

For an eBook on Amazon, you get a 70% royalty of sales if it's priced between $2.99 and $9.99. For anything less than $2.99 or over $9.99, that percentage goes down to 35%. Again, this is still a lot higher than what'd you get with traditional publishing, but you should generally still price within that 70% royalty range.

This isn't so that you can get the most money, though. As I've said, sales are not the real goal here. The goal here is to get as many *readers* as possible, and pricing your book too low can actually keep you from doing that.

You might think that you'll get more readers by making your book cheaper, but that's not the case here. People are going to gauge the potential quality of your book by what you provide (title, cover, description, first few pages, etcetera), what others provide (reviews and rank), *and* the price. While an unreasonably high price is going to deter most readers, an exceptionally low one is going to make them suspicious.

Remember, a potential reader will be investing both their money and their time into the book. A reasonable price will indicate that the book is worth the time that they'll then have to spend to get the contents of the book. If it's priced otherwise, then it might mean that it'll end up being a waste of time. An overly cheap book usually leads people to decide that the book is probably commensurate with the price, and thus not worth even the dollar that it does cost.

There's only one exception to this, and that's a marketing strategy that we'll discuss later on. The general rule of thumb is to price your eBook *typically*, or around what most books like it costs. That can differ quite a bit based on its exact categories, length, etcetera, but this range is usually around $5 to $8.

An eBook is the best format to reach the most readers, but having both digital and physical versions of your book available will reach both of the main types of readers. People are still reading old-school, so let's talk about how to create your paperback.

Paperback

I'll admit it—there is something to a physical copy of your book. Getting to hold your book in your hands is the single greatest tangible affirmation of all the work that you've done. Seeing your words printed on paper is a thrill. But creating a paperback version of your book serves more than just your ego.

A paperback copy of our book made by KDP's Print-on-Demand service.

A paperback version is just as valuable as a digital one for a few reasons. First, a physical version of your book will most contribute to its overall "legitimacy" and best establish your credibility as an author. While, again, an eBook is just as legitimate and valuable as a physical book and will help with your legitimacy, there's something about paper and ink that people value and trust, and that you can capitalize on to maximize your image as a through-and-through "author" authority.

And although it's true that a lot of people have shifted to using Kindles and other e-reader devices, a solid section of the market still remains firmly devoted to buying and reading physical copies. As I've already said, the biggest thing that you want to do with publishing your book is to make your book as available and as accessible as possible. An eBook will reach a lot of readers, but both versions will reach (almost) *all* readers.

Creating a paperback version of your book is no longer the difficult and tedious process that it had used to be. Amazon offers a Print-On-Demand (**POD**) service that empowers the independent author to be able to create physical copies of their book without any of the usual upfront costs.

Print-On-Demand works exactly how you probably think it does. Instead of pre-producing a set number of copies in case they sell, Amazon's **POD** literally makes and ships out your paperback copies as they are purchased. By creating copies only when needed, they are able to offer the paperback feature to all authors with a book that's with a long enough page length to be able to bind. And from what I've seen, as long as your book is more than the length of a short story, you will be able to make a paperback.

Setting up your paperback is just as quick and easy as setting up your eBook. You can use the information that you've already done for the ISBN and eBook to fill in for the paperback, taking care to change what's needed. This means making sure the ISBN, cover, text file, and price is specific to the paperback.

Pricing is probably one of the most noted differences between the eBook and the paperback. With the paperback, you're entitled to 60% of royalties in almost all marketplaces no matter what you set your price at. However, that 60% is after the cost of production. Unlike an eBook, a paperback obviously requires physical material and labor to make and deliver it to your reader. Amazon calculates that cost based on book length, specified dimensions, gloss versus matte cover, type of paper desired, etcetera, so this number can vary quite a bit depending on the book.

The minimum price that you can set your paperback at is based on Amazon's calculated cost of production, so that at least they can't lose money by making your book. From there, Amazon can show you what you'd be receiving for each sale after factoring the cost of production.

Like the eBook, you don't want to price your paperback too low. From what I've noticed, paperbacks priced under $6 are an obvious "giveaway" of being independently published, and I think that a lot of people may be able to notice that as well. You should try to price your paperback like your eBook—again, *typically*. Base your price point around what books similar to yours are selling at.

With the eBook, you're able to preview how it looks before ultimately submitting it. Since a paperback isn't a digital product, you won't be able to really see what it looks like online. That's why Amazon allows you to get a beta copy of your paperback to ensure that everything is right prior to releasing it publicly.

I highly, highly recommend doing this. For just the cost of production, you can check over your book from cover to cover. You really want to take advantage of this one last chance to catch any mistakes. Because trust me, if you don't, your readers definitely will.

Once you can see for yourself that everything is printed correctly, you're ready to release your paperback to the world.

Audiobook

Having both the eBook and paperback versions of your book available allows you to access the two main reading markets. But there's a fast-growing subset of readers who don't do it with their eyes. The savvier independent authors know to reach them. If you want to get your book to as many people as possible, then you absolutely have to consider making an audiobook version.

Audiobooks are quickly taking a bigger and bigger share of the reading market, and for good reason. I mean, think about it. Most people these days are way too busy to be able to sit down and keep their eyes on a page or screen. Reading has become a leisure activity that people have to work to fit somewhere into their schedules. The ability to consume a book during the daily commute or while working out is an incredible advantage of the audiobook that no other format can offer.

Since Amazon has acquired Audible, it's now just as easy to create the audiobook as the eBook and paperback. You'll first have to create an account on Audiobook Creation Exchange (ACX), which is linked directly to your KDP account. Just like KDP, ACX will walk you through the entire process.

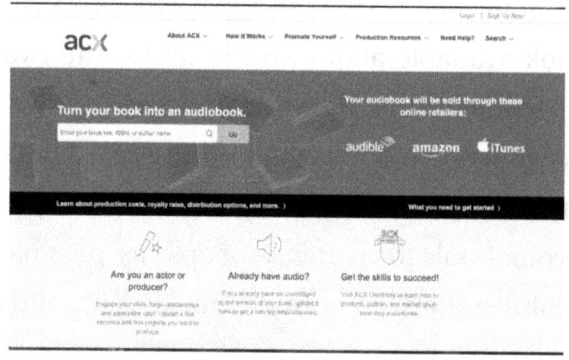

ACX's website.

Royalties for an audiobook are slightly lower, but still incredibly worthwhile. By producing it exclusively through ACX (which is the only worthwhile audiobook platform anyhow), you're entitled to 40% of audiobook royalties on Audible, Amazon, and iTunes.

The main difference between creating an audiobook and an eBook/paperback is obviously that it's not just an upload-and-submit process. You have to first find a narrator to create the actual audio format.

Luckily, there is no shortage of narrators (also called "producers) available on ACX. ACX allows you to post an audition with a description of the book, specifications for the exact type of voice desired, and a short sample of the book for those that are interested in the project to use to audition. From there, you'll receive and review these audio samples to ultimately decide which narrator best fits the book.

ACX explains the production process right on their website.

The voice that you're looking for to narrate your book doesn't necessarily always mean a better, professional version of your own. You might consider your reading audience and who they might be, or what kind of person would give the best "tone" of your book. Ultimately I recommend that you keep your options pretty open, avoiding any overly distracting accents or cadences, and look for someone that can best translate the way your book "sounds" on the page to literally in your ears. That's kind of an esoteric thing to say, but you'll totally get what I'm saying once you're in this process for yourself. When you eventually hear "the one", it's all going to click.

And once you find the one, you can then send that narrator an offer for the project. This offer will include the timeline of when you want it done and either an offer to split the royalties (good if you have a low production budget) or a set pay per finished hour (best if you want to attract high-quality narrators). If they accept your offer, then the production is handed over to them to do their narrating magic. At that point, you basically just wait until they're done with their end to approve the final production and then for ACX to ultimately approve your book to go live.

I do need to mention here that it can take up to six (that's right, six!) weeks for ACX to complete their audio review. With that being considered, it's especially important that you start the audiobook as soon as the eBook is live on Amazon. You obviously want to get every format out as soon as you can and the audiobook takes the longest by far. Like I have said, it's not hard, but it does take some considerable time and you need to account for that.

Hardback

With an audiobook, eBook, and paperback format available for your book, you'll be able to reach each and every type of reader out there. However, that's not *every* single possible format of a book that's out there. If you're a particularly observant type of person, then you might have already noticed that I haven't yet talked about creating a hardback. And that's not by accident, obviously. Making a hardback version of your book available is often a lot more trouble than it's worth.

A paperback is easy to make through KDP and will cover the people that prefer a book on the old school ink and paper. Very, very few readers are going to be deterred by a book's soft cover or lack of a dust jacket, and so having a hardback is not necessarily going to open up access to another type of reader. The process is also a whole lot more complicated to complete than any other format. As of early 2021, KDP has started offering a hardback option to specially invited authors. However, it's still in beta mode. What we've seen is that it's difficult to figure out the template for the cover, since they currently don't have a template generator, and they only offer a case jacket laminate (which, to give you some reference, looks like a textbook). It's cumbersome, time-consuming, and doesn't really bring in more readers.

However, a hardback certainly isn't going to *hurt* you. A hardback is more or less about the author's own ego and getting to see their book in the most "professional" format available. And admittedly, that can be a very attractive thing for some people. Hell, that's at least part of why I've done them myself. A hardback may also help further legitimize your book and contribute to that professionally-published product and image. If you want a hardback, then by all means do it, but understand that it's a lot trickier to accomplish than any other format and, unlike any other format, it's probably not going to be worth the work.

But if that still doesn't deter you, then here's the work.

If you are alright with KDP's only hardback option (which, again, just looks like a small textbook and not like the most typical hardback book), then you'll just need to work with your designer to get your current cover design to work. From our experience, it's a lot of trial and error to tweak the dimensions.

If you have your heart set on an "official" hardback with a dust jacket, you're going to need to find a third-party POD service to create your hardback on and then later link up to your Amazon. In my experience, IngramSpark is the best one. You'll have to create an account and enter in your book's information. A lot of it can be copied directly from the paperback such as the description, key words, and so on, but remember that you'll need to register a new ISBN and other details specific to this format.

IngramSpark has a very user-friendly design.

If you're going to keep the dimensions of the hardback the same as the dimensions of the paperback, then you're probably fine to use the same formatted PDF for the interior text. For the cover, however, you will at least need to change the barcode on the back. If you've decided to have a dust jacket, then you're probably going to need to go back to your cover designer and see if they can do this for you and, if so, for how much.

The flaps of the dust jacket typically have the book's description and the author's bio. So, you'll need to move the description off of the back cover and either replace it with blurbs about the book or especially glowing reviews, or to make it entirely blank. Again, this is something that you are not going to be able to manage on your own, so I'd strongly recommend getting a cover designer that knows how to do this before you even start the process to set up the hardback.

Once you have all of this done, you'll need to claim your hardback on Amazon. With IngramSpark, the hardback version will be available directly on your book's page alongside all of the other formats.

So with or without the hardback, you'll have your book available to anyone and everyone that wants to read it. Now, let's talk about how to get people to *want* to read it.

Launching Your Book

You can't simply hit "publish" on your book and get an overnight success. Those kinds of runaway bestsellers are incredibly rare and not something that you should even remotely expect. You might've written a work of genius, sure, but merit alone is not going to get readers. If you want people to know about your book, you have to *tell* them.

And if you don't put any thought or effort into how you're going to get the word out, then everything else you've done was basically for nothing. A book launch is critical to raising your visibility, attracting interest, and getting readers.

A book launch is essentially a plan of marketing strategies coordinated around your book's publication date. You need to start working on your book launch as soon as you start writing the book itself. It really is that important. By starting to build a base of potential readers before the book is done, you give yourself a lot more time to grow this base and stand a greater chance of converting more people over to actual readers. This is a positive snowball effect that gets you more sales, reviews, attention, everything that feeds into the possibility of achieving that coveted bestseller status.

A book launch isn't just one action of advertising. Think of it as a machine built up of individual moving parts. A successful book launch has multiple components that work simultaneously. That's also why it's so important to start this process as early as possible. The more time that you have to set each part up, the more effectively they'll come together toward that ultimate goal of promoting your book.

Online Presence

We'll start talking about marketing by talking about online presence. The way it works these days is that if you don't have some kind of presence in the online world, then you basically have nothing. If you want to generate any interest in your book and get the word out, then you *need* to get online. There's no other way about it.

At the absolute minimum, a halfway professional author will have a dedicated author's website and regularly update their site with blog posts to promote their book. Even if you know nothing about site-building or web design, you can still create a pretty decent site by using a service like Squarespace or Wix. These site-building sites exist specifically for those that don't know how to do this, and thus are incredibly user-friendly and will walk you through the entire setup.

The first part of building a website is picking out a domain name. Most domain names are some variant of the author's name. This makes sense, as the author (i.e. you) is the whole attraction here. I recommend claiming a domain name as absolutely close to your name (or, if you're using a pen name, the nom de plume) as possible. If it's already taken, you can sometimes contact the current domain owner and be able to buy the name off of them.

If this isn't possible, you could do with adding "author" or something related to your book(s) with your name. Generally, you want to keep the domain short and sweet (and thus easy to remember or type in), so pick and choose your characters with care.

With that out of the way comes the actual site. If you're going to do this yourself through a DIY website builder, then there's a couple basics of web design to consider. The ultimate goal here is to create a website that's pleasant to look at, presents its information clearly, and is easy to navigate. Let's go through all three.

You don't necessarily have to be a designer to make something look good. The easiest and often quickest way to achieve at least a decent aesthetic on your website is by taking a "less is more" approach and adopting the minimalist mindset. It's really hard to mess this up if you're keeping things simple. That means no complicated graphics, weird fonts, or colors from all over the rainbow. Look for readable text, a two-color palette, clean visuals, etcetera. And whichever design elements you choose for one page of your website has to be applied to *all* of them—consistency is key.

The homepage to my website. I'm not saying it's perfect, but it's an example of how keeping things simple will result in a clean and polished final look.

The reason people will visit your website is because they have had their interest piqued and want to learn more. So, you need to display the most important information first and foremost. What this usually means for an author's website is that you need your homepage to display your latest release. Ideally, this will look like an image of the front cover alongside a condensed version of your sales pitch (quick review: that's what the book is about, what they'll get from reading it, and why you're qualified to write it) and a link to every format available for purchase. The easier it is for them to get this information about your book, the more likely they will go so far as to actually read it.

The key to making your information easy to get is by making a website that's easy to navigate. After the aforementioned homepage, an author's website is usually going to also have these pages: "books", "about", and "contact". The "books" page is to list, describe, and link any and all books (or other relevant media) that you have published. The "about" page is more specifically about *you*, and so it's just a more detailed biography. And just as obviously, your contact page should have all of your professional social media accounts (yeah, you'll need those too) but *not* list your email (unless you enjoy getting slammed with spam). Instead, you can set up a contact form that visitors can send messages through.

Aside from a contact form, I also highly recommend creating an emailing that visitors can sign up for through your site. You can use SurveyMonkey, Gmail, or some other data-compiling service to build the email list. However, only add those that have asked to be put on them and only email them what's worth sharing (most recommended is promotional articles related to your book release).

You may also want to consider setting up a specific page on your website to put up blog posts. I'll be detailing exactly what kind of blog posts that you want to be writing and uploading a little later on, but for now know that they can be very well worth your time. Regularly adding valuable content related to your book(s) will increase traffic to your site, earn you repeat visitors, and boost that search engine optimization (SEO) we had mentioned earlier.

Basically, SEO refers to how the algorithms for search engines like Google calculate what web content gets ranked in the results. The more quality content you have on your site, the higher up your website will be displayed in related search results. This means more visibility, which means more traffic to your site, which means more sales, you see where we're going? Hopefully you get the idea by now.

Ok, back to site navigation. To make your website as easy to navigate as possible, I highly recommend adding both a navigation menu at the top of each webpage *and* directly on the home screen. This is the best way to make your website so clearly and obviously mapped out that even your technologically illiterate great-aunt could find where to go and buy your book off of it. Also, you have to make sure that your site is mobile-friendly. Most people are constantly on the go and most likely to be browsing on their phone or tablet. A site that doesn't work on anything but a desktop is going to send away many, if not most, of your potential site visitors.

It all sounds like a lot but, like I've said, it's pretty doable through a site-building service. But you do need to keep in mind that platforms like Squarespace will typically produce websites with pretty clear templates. So, if someone is familiar with what these drag-and-drop websites look like, they're going to be able to tell that you did not get your website professionally designed. And that's not necessarily a *bad* thing. If everything on your site is done cleanly and stylishly, this is definitely not going to be a mark against your professionalism.

However, if you do want to give off that extra (perceived) air of legitimacy, you may want to consider hiring a professional website designer. If you decide to go ahead with the professional route, you then need to consider *which* you want to work with: a freelancer or a design agency.

If you have a lower production budget and/or want to work most directly with someone, then a freelance website designer is probably going to be the best option. An individual is typically going to be less expensive than hiring the services of an agency, and the process will usually be a bit more personally tailored to you, but they may take longer to complete the project and the quality isn't always guaranteed.

As with anyone else, you always want to check their portfolios and reviews before committing to any one designer. Freelance platforms like Upwork will usually include their past projects and client reviews so that you can best vet the people listed. Whether you go for an individual or an agency, you should always do some independent research to verify their work anyhow.

A design agency is almost always going to be more expensive than a freelancer, so this means that this option is usually reserved only for those with a high production budget. Aside from price, you may want to choose an agency if you're looking to get an established, streamlined service that almost always offers a money-back guarantee of your satisfaction with what they will deliver.

If you're looking for something very specific, I again recommend my reverse-search method: if you find a site that you particularly like or is in line with what you want, you can always try to contact the site owner to see who built the site and reach out to the designer from there.

Whether you go with an individual or an agency, always make sure that they can make the site editable so that you can always go in to change and update information as needed, whenever you need to. Should you follow my advice and create a blog, you'll also want to make sure that you can either regularly upload on that page or have a link to your blog clearly displayed on your website.

Back to the blogging thing—you really should do it. For all the reasons I've said, it's definitely worth your time. And it doesn't have to take that much of it, either. I have a little shortcut to creating blog content without necessarily creating *new* content.

My trick is to take a section of the book that I'm currently promoting and repurpose it into a blog post. I'll do a little editing and move some stuff around until it's its own self-contained article that feels complete. From there, all I have to do is throw a catchy title at the top and plug the book that it's from at the bottom, and then it's totally good to go live. And that's it. With this method, you'll spend probably less than an hour on each blog post to generate that much more traffic to your site and interest in your book.

Even if it took longer, any time spent on promotional endeavors is almost always worth it. One endeavor that always *is* worth it is joining social media. But your personal Instagram page isn't going to cut it here. You're going to need to create professional accounts that work toward your book and brand.

I'm sure you're already aware of social media and its prevalence. Unless you're an intentional hermit, you probably have a Facebook account at the very minimum. The fact of today is that just about everybody is on everything and if you want to maximize your reach and audience, then that means that you have to get on it too.

For authors, I've noticed that Twitter and Facebook are the best platforms to focus on. If your book is centered on or written for a specific industry, then you may need to figure out which platform that your industry uses the most and then focus on that one. Either way, you should join every big platform possible—LinkedIn, Twitter, Instagram, YouTube, Facebook, Pinterest, even TikTok if it's still around. I honestly recommend setting up an account on everything even if you have no intention of ever posting on there. At the very least, you'll be able to claim the same username on everything and deter fake accounts from popping up under your name (yes, this does actually happen).

Like your website's domain name, your username should be something that's as close to your name as possible. Your username should also ideally be the same across all platforms, so always check the availability on *everything* before you choose one. Having the same username on everything not only helps enforce your brand, but it also makes it much easier for your followers on one platform to find you on another. And the easier it is to find you means the more likely it is that you'll get followers and grow your audience.

To that end, I also recommend having the same profile picture across all accounts. Someone who's never seen you in person may not recognize you in different pictures, but they definitely will recognize *that* specific picture. Pro tip: the author picture that you've already taken for your book will work perfectly for your profiles. By using the photo that your readers have already seen, you get to save some time and build your brand all in one move.

Most platforms will allow you to include a link in your bio. While it may seem like a smart move to include a link to your other core platform there, I actually recommend saving that one link space to direct your followers to your Amazon page. After all, the whole point is to ultimately get them there. That's why having the same usernames and profile pictures are especially important here.

Consistency across your social media accounts is absolutely key to building your brand online. Choosing and sticking to the same usernames, profile pictures, style, and tone of your posts all work to create and display one curated image of yourself. Conversely, having wildly different accounts is only going to confuse your followers. You need to have one single clear message of who you are and what you do. If you display one persona on a platform and a different one on another, then nobody is going to know who you really are. And if your audience can't get to really know you, then it's going to be impossible for them to connect to, engage with, or (frankly) care about what you're trying to get out to them.

You can't just be content with empty accounts under your name, of course. If you expect to gain followers and grow an audience, then you have to start getting active online. There's two main ways to be active on social media: engaging with others online and actually creating content.

Engaging is the easy part. Start by following accounts similar to yours or to what you want yours to be. Like their posts, share, and contribute appropriate comments. Supporting other people's content is not only a nice thing to do and generates karma, but it's also just smart. Platform algorithms pick up on interactive accounts to promote to others. So, it's a win-win all around.

Algorithms also like accounts that consistently create content. Content just means what you post—videos, pictures, text, or any other media. What you post can be anything you want, so long as it's relevant to your book and/or profession and is appropriate. Now, what's considered "appropriate" may differ based on the person, but I'd go by "what will not harm your career". Someone in a more casual business (or, like me, no longer really gives a shit) may decide that using expletives is part of their brand and won't lose them any business, but a schoolteacher would find themselves in a world of trouble. It all depends on who you are, how you make your money, and what you want your brand to say about you.

So depending on the details of your specific situation, your definition of "appropriate" may be very different than someone else's. That being said, what you deem to be appropriate and relevant to your brand almost never really means divulging the details of your personal life or your beliefs. That kind of stuff should be kept on a personal, *private* account—this means checking privacy settings on all of your personal accounts and limiting who has access to your content.

Let me explain. Yes, I believe that everyone is entitled to the freedom of expression and I fully support exercising that right. However, if you are going to do that on a professional account, you need to at least understand that you can severely impact your ability to get readers and do business with others.

Aside from that, they're almost always irrelevant to your brand. Unless your book is quite literally about your religious or political beliefs, they really don't have any actual place in your posts. *Relevance* is the key to determining whether it really needs to be posted or if it should be left out.

What you post also needs to be considered of quality. How can you decide if your content is high or low quality? Consider if you put any time, thought, and effort into making it. If so, you're probably fine. If it's some industry meme you'd screenshotted from another account, that's not worth putting up (and also plagiarism). You need to be posting consistently, but spamming your followers with stuff that's not worth being on their feed is going to lead to losing your audience entirely.

Let's review: you want to post high-quality content on a regular basis. Beyond that, you also want to try to create content that gets your followers to further *interact* with what you post. A high interaction will raise your visibility, which will then grow your audience base. I've got a couple of tricks stolen straight from career influencers that will get your followers to like, comment, and share as much as possible:

- Include a question in your caption or post and invite your followers to answer it in the comments.
- Post daily polls related to your book and/or industry.
- Create a poll by asking your followers to either "like" the post if they prefer x or to comment on the post if they prefer y. Doing either one will create interaction with the post.
- Ask your followers to specifically "like" a post. This typically works best at the end of a video.
- Run a giveaway. For the author's account, that means a book giveaway. A book giveaway is a really strategic move that promotes the book *and* grows your following on social media. The best way to do this is to post a picture of the book and to announce the giveaway in the caption and explain the rules of it. Typically, entrance requirements will include: to follow your account, to like the picture, and sometimes to also tag someone in the comments. From there, you may need to get some kind of software to catalogue all of the

qualifying accounts and choose the winners at random. This also means that you'll have to either pay for copies of the book to ship out to the winners or to email the eBook out and hope that they don't share it.
- Consider collaborating with other similar accounts. This is a mutually beneficial endeavor that creates content for both people and puts it in front of two different audiences.

So, regularly uploading quality posts that invite further engagement from your followers is going to raise your account's visibility, which will earn you followers, which will grow your audience. And so when you make a few incredibly selective posts to promote your book, it's more than likely that at least a few of your followers will convert to actual readers. The strategic use of social media is essentially free advertising along with the long-lasting benefits of building an audience.

It's pretty likely that some of your followers will read your book. However, I have a guaranteed way to get readers, reviews, and attention for your book. If you're already open to the idea of giveaways on social media, then you'll definitely be interested in the promotional power of the ARC.

ARCs

When your book first goes live for purchase, the only thing that people have to judge its quality by is what you provide in the description and preview pages. If you did it right, that should all be good enough to win the reader over. However, people are most likely to trust the word of other people.

How often do you read reviews? Probably every time you browse. And they often become the deciding factor in your ultimate decision. As they should—a previous customer has written it for a reason. A high-rated product may not really be the best, but you have probably at least been steered away from wasting your money by reading reviews that were contrary to what was advertised. Reading reviews is one aspect of being a responsible consumer and making good purchases.

The most surefire way to get it is by giving out Advanced Reader Copies, or ARCs, to a select group of people. With an ARC, the understanding is that the person is given a free copy of your book before it is published with the expectation (but not the requirement) that they will write and post a public review of the book once it is published. In this arrangement, they get to read the book for free, and you get a review.

The five-star rating acts as the general public consensus of the quality of any offered product and the reviews are only the details. A high-star rating catches your eye, whereas a low or no-star rating will make you immediately scroll away. For your book, you obviously want good reviews and all the stars that you can get. I don't even need to explain why. I mean, duh, anybody would want that.

You might wonder, why would you give a copy away for a review when you might get one from an actual sale? Well, to get what is often called an "organic" review, you'll have to wait for someone to buy and read the whole thing first. That could take some time, and there's no guarantee that you'll actually get one. When it comes to ranking in your book's categories, time is your enemy. If you're looking to get any visibility for your book, then you cannot rely on waiting for reviews to come in organically.

ARC readers all know the implicit expectation of writing a review and they almost always actually do it. And while you absolutely cannot explicitly ask for a review, much less ask for a five-star one, these readers will typically leave a positive review. These initial, often positive reviews are incredibly important to get your book ranked and picked up by the algorithm.

Amazon's algorithm looks for products with traction. The biggest signal that a book has traction is partly sales, but reviews (both good and bad) are what often prompts the algorithm the most. Once this is picked up by the algorithm, your book will then start to rank higher and higher within its categories and may even get pushed as a "recommended" item to potential readers. This all creates more sales, which creates more reviews, which creates higher ranking...you get the idea.

But to get that ball rolling, you have to give it a little artificial push through ARCs. It's rare, if not just about impossible, for a book to get that kind of attention without any strategic measures like this. And you're a first-time author. You have no prior reading audience to rely on. If you really want to ensure that your book gets noticed, then you need to be giving out those ARCs.

While advance audiobooks do exist, I recommend either choosing between giving out advance paperbacks or advance eBooks. As a paperback will have to be physically produced and shipped out, this is obviously going to be the more expensive option. Sometimes, though, the extra money spent could pay off.

That depends on *who* you're trying to get to review your book. For example, if you're looking to give all of your followers an ARC, then an eBook is probably the most economically sensible option. If you're trying to get a specific person, like someone famous in your field or an industry expert, then shipping a physical copy to their office or house is generally going to be a lot more impressive to them and could increase the likelihood that they'll actually read and review your book.

So, my general advice with choosing between print and digital ARCs is to consider your production budget and the goal for your ARC reach. Print is good to get a specific person whose review would hold a lot of weight, while digital is ultimately best if you're just trying to get it out to anyone who asks for it and to play the numbers game.

At this point, you've probably already noticed that Amazon's Print-On-Demand service really does mean that—they only produce books as they're sold, and they limit the number of beta copies that an author can purchase. So if you're looking to print up and ship out any number of print books, you'll have to do that somewhere else.

This is where, like the issue with the hardbacks, IngramSpark comes in. You can set up and order however many copies of your book as you'd like through their platform. All of the information and files that you uploaded to Amazon can be copied and pasted directly to their site. However, I'd add a watermark on the cover to distinguish it as an Advanced Reader Copy. You literally can create an obnoxiously big banner over the front and back cover that says "ADVANCED READER COPY—NOT FOR SALE". It's unlikely that anybody would turn around and try to sell your copy, but it's a low-effort measure to ensure that they won't be able to.

Before ordering all of your ARCs, I strongly recommend first getting just one as a quality check. If there is some unexpected error with the printing, it's a lot better to have only one messed-up book rather than opening up a whole caseload of them (not to mention wasting all of that time and money as well).

Once you've gotten your beta copy and have made sure that everything looks completely good, you can then place your entire order. I calculate how many print ARCs that I need to order by adding up the sum of how many worthwhile people asked for one and the number of specific people I'm trying to get to review the book. Then, I add a couple more to that number just to be safe.

But that's just my personal method. Being a professional, my production budget is on the higher end and I'd personally always rather have too many than not enough. You may prefer to order an exact number, and there's absolutely nothing wrong with that. As they say, YMMV (your mileage may vary).

From here, it's just a matter of shipping the copies out to all of your advanced readers. If you're trying to get a specific person to read your book (who may not have necessarily asked for it), I recommend sending them a copy through a FedEx package. They have to sign for the package, so it guarantees that they'll receive it and get their attention enough to at least open the package.

Now, that doesn't mean they'll actually read it, but they're likely to skim the back cover at the very least. Again, if you did it all right, that should be enough to get them to open the book. I also recommend sticking a personalized note in the book itself to thank them for their time and include your contact information. If they like the book, they will probably contact you. If you're a particularly persistent person, then you might even want to try (politely) following up with them directly to ask if they read your book and whether they'd be willing to write a blurb or review for it.

If you're looking to just send out your ARCs digitally, then this process is a hell of a lot quicker (not to mention cheaper). It's just a matter of sending a watermarked PDF file to the email addresses of everyone who asked to read it.

Whether you decide to send out your ARCs physically, digitally, or both, you should always make sure to get your reader's email address. By having their contact information, you can create and send an automated email to gently remind your ARC recipients to write a review once your book is live. Make sure to thank them for reading before noting that they can now write a review.

You can most easily accomplish this by sending out one mass email through an automation service like Mailchimp. I personally start by writing a subject line that will just about guarantee a high open rate on my emails. To best get someone's attention, you want to get their name in the subject line. A personalized message is an effective message. That's why you also need to directly address them and make a call to action. Here's an example of how to email out your ARC:

Hello [name of person you're emailing],

Thanks to all of your encouragement, I've finally finished writing my book. [Title of book] will be available on Amazon on [set date of publication], and I'd really like to get as many readers as possible.

I've attached an Advanced Reader Copy of my book to this email. It would absolutely mean the world to me if you read my book. Please take a look and let me know what you think—if it was good, bad, just okay, anything. I really appreciate your support.

Thank you,

[your name]

Once your book actually goes live, you're going to want to send out something along the lines of this:

Hello [name of person],

Thank you again for reading my book. [Title of book] is finally now available on Amazon, and I need to ask for one more favor.

If you can, I'd really appreciate it if you would write an honest review of my book. It wouldn't take much more than another few minutes of your time and it would greatly help my book get more readers. The link to my Amazon page is attached below. Again, thank you so much.

Sincerely,

[your name]

You don't want to send that email out to everyone at once, though. You need to stagger when and how many reviews come in, or else Amazon might flag the incoming reviews as being paid for or otherwise ill-gotten and remove them entirely. To avoid this, you should separate the email list out to different groups of people to contact on different days. That way, the reviews don't start rolling in all at once and Amazon's algorithm won't have any potential issues to pick up on.

It also wouldn't hurt to ask them to explicitly say that they were given a book to review. Amazon allows ARC reviews, but being upfront about them best ensures that the reviews will actually get to stay up. Quick review: to avoid any and all issues, always make sure that you divide your contact list, stagger your reminder email, don't outright ask for positive reviews, and encourage your readers to be as honest as possible regarding their opinion of the book and how they got their copy.

ARC reviews usually end up being a great way to kick off the book's release and raise its visibility. With ARCs, you may also be able to retain and grow a readership base. You're already sending out emails, so why not consider creating a newsletter?

Newsletter

If you go to most author websites, you're going to immediately notice a banner or pop-up ad inviting you to sign up for their newsletter. This is one of the most common, easy, and effective ways to grow your audience and overall reach.

A newsletter is beneficial to both you and your subscribers. They get valuable information and/or an entertaining short read that they actually want delivered straight to their email, and you get an opportunity to interest them in your forthcoming book. It's a win-win all around.

The best way to get newsletter subscribers is to offer them a free article or chapter of your book in exchange for entering their email. Since they feel like they're immediately benefiting, they'll be much more likely to join. Offering a free sample or preview of your book will obviously also help sell the rest of it, too.

Writing and sending out a regular newsletter is surprisingly not that much more time and work. In fact, you really shouldn't send out your newsletter too frequently. Less is actually more here. Oftentimes you can even repurpose parts of your book the same way that you would a blog post into a newsletter.

Just make sure that what you end up sending out is high-quality, conducive to your brand, and truly worth your reader's time. Never take their attention for granted. Nobody likes to be treated only as a potential dollar. You need to build up their trust by sending free, worthwhile information before you blatantly advertise your book.

When you do decide to promote your book through your newsletter, make it so that it's another perceived "good deal" for them. I highly recommend setting a temporary lower price for your book and promote it as a special sale made specifically for them. That way, they think that they're getting let in on an exclusive promotion and are more likely to click the link to your Amazon page and buy it.

No matter what, your readers need to feel like they benefit from your newsletter. If you're putting low-quality material or obvious ads in their inbox, they will eventually unsubscribe. But if you respect your readers and are willing to write a little extra, then you'll earn book-buyers for life.

Ad Campaigns

Like any other product, there's an age-old way to get some nearly surefire attention: advertising. But before we get into talking about spending more money, let's go back to the production budget. Aside from coming up with an acceptable monetary number that you're willing to put toward your book, you also need to figure out exactly *what* you want to spend it on. Each and every possible expense should be considered and calculated, and advertising is included. And if one of your goals for the book actually is sales, then advertising will especially pay off for you.

Let me explain. If you do it right, then the money spent running ads will often directly translate back into money earned through book sales. You might end up spending less than a $1 per paying reader that you otherwise wouldn't get, and that's pretty obviously worth it. It might cut into your overall profits, but it'll also drive up your sales, which means your ranking, etcetera... again, worth it.

Okay, but say you don't actually make your money back through immediate sales. For example, let's suppose that you spend $50 on a small ad campaign. You track your traffic and see that that ad got you two book sales. After Amazon's cut, you might be getting like $10. So, you're out $40 *right now*, but you might've just gotten two new dedicated readers. They might go on to buy your next book and any book after that. A lifelong reader becomes part of your built-in audience that are more inclined to write reviews, interact with your social media posts, and even recommend your book to other people. I'd personally buy that kind of loyalty at any price, but they pay themselves off.

And if you've been smart enough to use your book to leverage attention towards another business venture, go on to become a client. Any service or product that you might have is probably going to cost more than your book. You might've spent a total of $20 to get them, but they may cut you checks for years to come. When you think about the typical customer acquisition cost (in business grad acronyms, CAC), that is remarkably low. So, again, very much worth it.

Like I've said, book sales shouldn't really be the big goal here anyhow. That's not where you're most likely to profit. Whatever you spend to get book sales can often lead to clients, which is where the real money is. That's what you can do with your book, and that's what an ad campaign can do for you.

The most effective resources to take out an ad campaign depend on your audience and your budget. Subreddits and Facebook groups related to your book or profession are always the most fail-safe option. You could also look to sponsor related content on a blog or digital media site, although the price on that can differ widely based on the size of their audience and perceived esteem of the site.

There are plenty of advertising avenues that you can take across a range of price points. You can advertise directly on Amazon, on specialized book sites like Goodreads and BookBub, on targeted Facebook groups, and more. It's difficult to direct where the definitive best place to advertise is because every book is different, but you can ultimately find it by considering where the people that are most likely to read your book are "hanging out" online and then go from there.

Since Amazon has a hold on arguably the largest part of the reading (or really any) market, they may be one of the easiest and most effective options for advertising.

You just need to go to the "Marketing" page and they'll walk you through how to set up an advertising account and create an ad campaign. Though before you do this, make sure that your book is actually eligible for advertising. For example, if your book has any expletives in the title (which I'm personally guilty of), you won't be able to advertise on Amazon. Pretty much any platform is going to have certain requirements like this, so you'd need to check that your book meets those standards anywhere.

But just because it costs a lot doesn't necessarily mean you're guaranteed to get sales. Your advertising has to be *effective* rather than just expensive. But sometimes money really is what makes it work, and paying for a well-crafted ad campaign can take your book to levels far beyond your imagination.

What Your Book Can Do For You

Like I've probably belabored at this point, you need to look at writing your non fiction book as an entrepreneurial endeavor. As with any entrepreneurial endeavor, you'll need to invest a great deal of time and work alongside some amount of money to create a book that is ultimately worth your efforts. And if you do it all right, your efforts are definitely going to pay off.

But I've said it before and I'll say it again: *you can not and should not expect your book to make you money directly through sales.* It's possible, but not likely, and you are going to set yourself up for failure if you adopt any other mindset. Fiction writers make their money off of royalties alone, but not writers like us. Our books are not the only thing we have. In fact, our books are really just to *enhance* what else we have.

If you can see the big picture, then you'll understand that finishing your book is only going to be the start. A book will position you as an expert in your field and earn you credibility like nothing else can. Point is, a big door is open now, and the only limit here is your imagination.

And if your imagination is a bit lacking, there's some things that just about *all* nonfiction authors can do with their book.

Getting Publicity

The biggest thing that you might not have thought of is that your book can get you quite literally free publicity. Ever thought about being on TV, going on a podcast, talking on a panel, being quoted in an article, or getting interviewed? Being an author makes you among the easiest sells for getting air time.

However, the press is not usually going to just come to you, especially not at first. You might have a book out now, but that alone isn't going to get the press to go to the trouble to find your information and contact you when they've already got a million pitches sitting in their inbox.

It's the same thing as getting readers; people aren't going to just automatically know you and your value without anyone telling them. You have to actively attract attention to actually get it.

The first step here is to create a good press kit. That's the first thing you'd show a potential publicity outlet, and that's the first thing they'd look at. Pitching without a press kit is like trying to drive across the country without gas. You've got a big goal, but you're not getting anywhere.

A press kit, also called a media kit, is basically a neat little package of material of all the information that the media might care to know and that you want publicized.

If you're not sure how to put it all together, there are plenty of templates available online that can guide you through it. And as always, learning by example is recommended. A lot of authors put their press kits right up on their website as both its own page and as a downloadable PDF. You can see what a similar author has done in their kit to get ideas for how to best create and shape yours.

The overall goal of a press kit is essentially to make it as easy as possible for a reporter, journalist, or anybody in the media to create and publicize a story. They won't have to do much work to research you because you're literally handing your information and even story suggestions to them on a silver platter. It's not that we're calling them lazy or anything. People working in the media are expected to turn out content on a constant basis. The less time and work that they have to put in to publish your story or incorporate you into one, the more likely they'll do it.

Your press kit should include:

- A high-resolution file of your professional headshot.
- Biography that includes any achievements, awards, etcetera.
- Any prior media coverage.
- Complete contact information.
- All of your (professional) social media.
- A complete summary of your book.
- The selling description of your book (the one already on the back cover will work).
- A list of the book's product details: full title, page count, publication date, ISBN, publisher, price, available formats, and where it's available to purchase.
- A full-resolution file of your book's cover.
- Selected reviews of your book.

- A list of books that are similar or comparable to your own.
- A short excerpt of your book.
- Suggested questions for an interview.

You might possibly need to include other things if they're specific to your book, field, other areas of work. Do keep in mind that you need to judge whether you're including something out of vanity or because it would make your kit more effective. It's pretty common for beginning authors to overload their press kits with information that doesn't actually matter (at least, in terms of getting air time or in articles).

Ultimately, a press kit is a quick and concise way to give a potential media outlet everything that they need to know about you and your book to write a story. And like I've said, the easier it is for them to write it, the more likely that they'll do it. Now, what if you wrote it for them?

A press release is basically a short, print-ready article that you've written that features you/your book without advertising it. It's an easy way for a publication to get material in exchange for plugging you/your book. But, how do you write a press release without obvious self-promotion?

Your goal here should be to craft an article that's useful to readers and hopefully relevant to the current news cycle. Send it in an email to a publication that you think it would best fit—local, regional, as broad as appropriate—including a nicely-written note and links to the book and your bio. Follow up if you haven't heard back but understand that this can be more or less a shot in the dark.

Let's be real here: you're sending yourself out unsolicited, and they're getting plenty of those every day. If you really want to give yourself the best chance of getting publicity, you may want to consider working with a PR firm. They know what they're doing and can oftentimes even guarantee some degree of press coverage.

Getting publicity also comes in many forms. To a certain extent, your published book will get you more publicity and attention for your business or personal brand. But it's important to start the process before the release date. That way, you build anticipation and curate your audience. One of the best ways to build a following is through speaking engagements.

Some people hear the term "speaking engagements" and fly into a mental frenzy—if public speaking isn't your favorite pastime, you can still find opportunities that will buoy your brand and book. The focus of the speaking event doesn't have to be solely about your book; it can cover anything in your area of expertise, though it is important to highlight the connection to your book and offer a glimpse at what the audience would get from reading it. Speaking engagements are too beneficial to be avoidable, but there are ways to make them more palatable.

Start locally in a town or city where you feel most comfortable and confident. It helps if the community there already knows your name or knows of your business. Bookstores and libraries are the obvious places to approach about doing a talk, and many are open to authors holding events since it boosts their businesses as well. Beyond that, seek out universities and colleges in your area. Oftentimes clubs have to pay for speakers to visit their schools, so if you offer them a good deal or even do it for free, you could land yourself the gig. Similarly, you can check out the local organizations that are relevant to your topic and would potentially be interested in having you as a speaker. People who are passionate about your subject will be more likely to commit the time and energy required to attend the event.

Don't be afraid to use social media to get speaking engagements, either. You can find events happening near you on Facebook and try to network that way.- Facebook live streams and other virtual events are ever more popular and regular in the post-pandemic world. These platforms allow you to reach a wider audience than what you could gather in person, removing several barriers to access which would otherwise prevent many from attending. You can promote your virtual speaking events via social media channels and gain publicity this way. These self-made virtual events allow you to drive the conversation and choose your approach, such as through an interview, a deep-dive into your topic, or the relevance to your audience. Cover any subject that you think represents you, your credentials, and the book's advantages, all leading into your brand.

The more assertive you are in seeking out opportunities, the larger the audience you'll reach. While the number of attendees at your speaking engagements won't equal books sold, this tactic will increase the exposure of you and your book, facilitating conversations about your topic of choice. If you have connections within your professional network, don't be afraid to reach out to them about opportunities to speak. The worst that can happen is that they say no, and rejection is inevitably going to happen in these publicity endeavors.

The harsh truth: gaining publicity takes a lot of time and effort, and you'll likely get a lot of rejections. But the speaking engagements you *do* get will boost your notoriety and contribute to awareness of your book, you, and your brand. No gig is too small to start with and, as you go, you'll find that speaking about your expertise and your book gets easier with practice. So even if you aren't the most outgoing person, the time and effort you put in will be well worth the results.

And once you've had some media attention or booked a speaking engagement, it's easier to get more. This accumulates until you're *known*, and it's obvious how you can leverage your relevance to getting business, jobs, and more.

Getting Business

And aside from these big things, a book can most directly help you get clients or to promote your business. What's the best way to show that you know what you're doing than by literally *writing the book* on your area of expertise? In some situations, the entire production cost of the book can even be written off as a "business expense" on your taxes. Of course, always check with an accountant to see if you can actually do this, but it's most often a totally legitimate write-off.

If you tend to work for others, then you'll find that your author status will make you stand out and put you miles ahead of other applicants. A book is a major credential that can help you get jobs previously out of reach.

Think about it. There are more people on this planet than ever before, and that means that there are often plenty of other qualified candidates that have already applied for any job that you might be looking at. The basic requirements for any given position keep rising. A college degree is the absolute least of it—employers want to see years of experience and expertise on resumes for an entry-level position, even for an internship. Your competition is highly educated, highly skilled, and *brutal*. In a sea of equally-good applicants, it can oftentimes be impossible to stand out.

Having your name on a front cover will put your application to the top. A book brands your abilities and expertise as *you*. A hiring manager is going to notice and remember someone that literally wrote the book on the same thing that they're applying to do.

Your Author Career

You don't have to stop at one book. You might not think you even had enough in you for *one* book, let alone two, but people with a specialized knowledge and/or skill set tend to have enough solid material to write an entire series. And if you've written this much, you might be able to write more.

The best way to tell if you really can write another book is to go back through all of the content presented in your book. Is there a chapter, section, even a sentence of something that you could expand on? If you can take a lesser-discussed idea presented in your first book and develop it into a complete thing of value, then you can write another book. And then from there...well, you can see where it'd go.

There are people who have built whole careers off of a book. Think about Tim Ferris. You probably know that name, but you probably don't know why you recognize it.

If you do know, though, then you've probably at least read his most notable book, *The 4-Hour Workweek*. This book was what gave him the public relevance and authority to write more best-selling books, get a short-lived TV show, and more. The book isn't necessarily what made him succeed as a startup advisor and investor, but you can't tell me that it didn't help at some level. And you definitely can't tell me that his book didn't give him the credit to get "lifestyle guru" listed on his Wikipedia page (or have a Wikipedia page at all).

There are a lot of Tim Ferrises in this world. There's also room for more. You can become *that* noteworthy person in your industry or field. A book is the best way to start to solidify yourself within that spot.

Afterword

Writing and publishing a book is one of the absolute best ways to build your brand. A book lets you reach your audience, show your expertise, and tell your story exactly how you want to. Being an author puts you ahead of everyone else and into a caliber of your own. Everybody has something to set them apart from the rest, but not many have *this*.

That's because writing a book is hard. If you've gotten this far, then you definitely understand all of the time and effort that it's going to take to get it done. It's planning out each and every part of what you're going to write and how you're going to write it. It's getting your butt in a chair and working at it for hours every single day. It's spending money on the right services to get

Afterword

your manuscript into an actual book. And it's doing even more work to get people to read it.

If knowing everything you have ahead of you doesn't deter you, if you have the time and energy to dedicate to this, and if you really commit to doing it, then *you can do this*. It's not impossible, obviously.

Yet you can have all the drive and desire in the world to write your book, but if you don't have the time to do it, then it's just not going to happen. After all, everyone only has 24 hours in their day. And if your time is already taken by other commitments, then you may want to consider looking for outside help.

I promise that this book isn't some sort of long-winded advertisement. If it was, you're all smart enough to have picked up on that from the first chapter. All I'm trying to do here is to show others how to do something that took me years to figure out for myself.

Remember, I didn't actually do it all on my own. I ended up needing someone else to help me finish my first book. While I surely had the expertise and the material in me, I had no way of getting it entirely down into a book. Again, writing is a specialized skill like any other. My co-author literally went to college to develop this skill.

Afterword

So, you could be like me. You might not be a writer or don't have the time to do it, and that's okay. It doesn't necessarily mean that you can't write a book.

Because here's a little secret: most of our kind of authors don't write their own books. They get a professional to do it for them And that makes perfect sense. We're entrepreneurs after all, not writers. You wouldn't expect a doctor to build their own furniture.

And again, this kind of book is an *investment*. You're putting in the time and money with the expectation of getting a worthwhile return. As we've covered, it's more than worth it if you do this right. But what if you don't?

You could throw all the time and money you have to make your book. You might get a perfect-looking product and advertise it to every possible reader out there, right. But if you're trusting that your novice, inexperienced attempt at writing will produce a coherent and quality book, you may end up wasting everything.

We've talked a lot about the benefits of getting a professional versus rolling the dice with the DIY route. You've already seen how just about every other step of the book process should be outsourced to someone

Afterword

who knows what they're doing. We've got experience and expertise with every step, including the writing.

For the busy entrepreneur, BrightRay Publishing is the most time and cost-effective choice. Instead of paying and waiting for each separate part of the book, we can give you a set timeline from the first draft to a published book in your hands for a set price.

And for those that already have the written part complete and are looking for the most efficient way to get it done, we also offer an all-in-one service to get everything else done. When you factor in the cost of getting each and every professional service separately, we work out to be a lot cheaper (not to mention easier). We've created a streamlined process to turn a manuscript into a book at a fraction of the time and none of the stress that it'd take for you to try to figure it out on your own.

This book outlines everything that you need to do to write, edit, publish, and promote your own book, and how to use it to build your brand. If you have the time and drive to follow everything here, you will be able to do it.

But if you're looking for other options, I've given you one. And if you want to learn more, give me about five

Afterword

minutes of your time to check out www.brightraypublishing.com. I might be able to save you months or even years of your life in the long run.

Glossary

Algorithm: The "computer brain" or process by which a site or platform determines how and what results are displayed. This is specifically referring to the algorithms behind the recommended items on Amazon and the search results on Google.

Amazon Best Seller: A product on Amazon that sells more than any other product in a category. This banner is awarded based only on sales, but it can be awarded for any category available.

Amazon Rank: The numerical rank shows how close (or far) a product is to the best seller in any category.

. . .

ARC: An initialism that stands for Advanced Reader Copy. An ARC is given out to either beta readers, critics, or to others in the hopes of generating feedback, interest, or reviews for the book prior to its official publication.

Beta Reader: Someone who reads an ARC specifically to give the author feedback, critique, and/or an overall impression of the book. A beta reader is often not a professional within the writing or publishing world. The ideal beta reader is someone within the book's target demographic.

Book Launch: A set of planned promotional strategies to market and advertise a book's publication. What strategies, and to what degree they are utilized, depends on the variances and goals of the book.

Brand: The story or image that you craft and control for your professional self or business. A brand is what helps clients/customers come to associate you/your company as *the* expert or business in that particular area.

<u>CAC</u>: An initialism that stands for Customer Acquisition Costs. In business terminology, this refers to how much one has to spend in order to get a customer.

<u>Categories</u>: Where a book is "placed" or what groups are most relevant and associated to the book in theme, topic, subject, etcetera. Categories determine the book's visibility and ability to rank on the distribution platform.

<u>Copyeditor</u>: A copyeditor reviews and revises a manuscript for its details on more of a micro basis. This means fact-checking, grammar, and other smaller parts of the manuscript to improve the read overall.

<u>Edit</u>: The process of rewriting and revising a manuscript until it is ready for publication.

. . .

<u>Editor</u>: An editor can look at a manuscript at any point, and so this term is relatively broad. An editor stands out here because it is typically the first professional to look over a first draft. They will usually take a macro view and suggest changes to the structure, tone, or other big things in the manuscript.

<u>First Draft</u>: The first completely-written version of a manuscript prior to any edits.

<u>Format</u>: The version that a book is produced or available in. This could mean paperback, hardback, digital, or audiobook.

<u>Freelancer</u>: A professional or a highly-skilled person that offers their services outside of regular employment. They will typically charge either an hourly rate or a flat fee for the project.

<u>Google Rank</u>: This rank specifically refers to the order that results are displayed in a Google search. Results are determined by relevance, authority of the source, and other factors.

• • •

Independent Publishing: The author directly publishes their work through a POD or digital distribution platform. They have to pay for the cost of their book's production and are not paid an advance, but they receive significantly higher royalties.

ISBN: An initialism that stands for International Standard Book Number. This is a 13-digit numerical code used for booksellers, publishers, and others to be able to identify a book. Independently-published books can register for an ISBN.

KDP Unlimited: A program available on KDP for independent authors. Books enrolled in KDP Unlimited can only be available on Amazon and can be read by all Kindle Unlimited subscribers in exchange for a payout based on what the program made each month and the author's individual KENP number.

KENP: Initialism that stands for Kindle Edition Normalized Pages. This is essentially the total number of pages that Kindle Unlimited members read of an author's books.

Keywords: Essentially the words and phrases tagged to a subject. These keywords determine the likelihood that the subject will appear in a search result.

Layout: The interior text design of a book.

Manuscript: The complete draft of a book prior to publication.

Mise en Page: In French, this means "to put on the page." This refers to the text layout or formatting.

Online Presence: You or your brand's digital visibility. This can be measured by followers on social media, Google rank, and other factors.

Organic Review: This is when someone buys a book without being personally asked to, reads it in its entirety, and then writes a completely honest review of it afterward.

Outsourcing: This refers to getting a freelancer or agency to do something for you rather than trying to do so yourself.

POD: An initialism that stands for Print on Demand. This refers to distribution networks that will produce a physical copy of a book only when one is sold. Printing books only when needed is what allows many independent authors to be able to publish their books in a physical format

Press Kit: A collection of documents, files, and other materials made readily accessible to a media outlet so that they can easily get information for publicity.

Press Release: A brief, print-ready story that features you or your book without blatant self-promotion.

. . .

Production Budget: The amount of money that an independent author has set to spend everything needed to publish and promote their book. This budget depends on what the author is willing to put up and the goals for the book.

Proofreader: A professional who will look over a manuscript only to detect and correct any technical issues with grammar and typos. They are one of the last professionals that will review a manuscript.

Rough Draft: Another term for a first draft.

Royalties: The percentage of the revenue that an author is entitled to. This percentage is typically a lot higher with independent publishing than in traditional publishing.

SEO: An initialism that stands for Search Engine Optimization. This essentially means the process of getting more visibility and attention online.

. . .

Target Audience: This is the most ideal reader for your book. A target audience does not necessarily have to be a certain demographic of people but rather a specific type of person.

Traditional Publishing: The author receives an advance for their book and does not have to put any money upfront into the production details of their book, but a publisher will keep a majority of the royalties and push the promotional responsibilities back onto the author.

Typesetting: Prior to digitalization, this was the process of physically arranging the text of a book so that it is ready for print.

Resources

GENERAL FREELANCE PLATFORMS

• Fiverr

Fiverr used to be famous for its $5 gigs, but you can now find freelancers at all price points. The platform has also faced some criticism for its treatment of freelancers, which could be expected from a brand that once advertised their bottom-barrel rates over the quality of work.

• Upwork

The other large freelancing platform. They advertise the quality of the talent listed on their site rather than the cost.

FOR WRITING AND PUBLISHING SERVICES

• Reedsy

A freelance platform where industry professionals can list their services. You can find freelancers here for every single step of the writing and publishing process at a variety of price points.

• Editorial Freelancers Association

All freelancers have to pay an annual due to be listed on this platform, so they are typically more "serious" than those on other sites. This platform offers both a directory of freelancers as well as a job list where you can post an ad and have those interested in the job contact you directly.

• Scribendi

This platform requires freelancers to apply and go through rigorous screening. This vetting of talent generally means a higher-quality end result.

FOR COVER DESIGNS

- 99designs

In our opinion, the absolute best way to get covers. This platform allows you to either connect directly with freelancers or to host a design contest. You can give feedback and specifications to tailor the contest entries until you get *exactly* what you want, which is ultimately the only one that you pay for.

FOR INTERIOR TEXT LAYOUT

- Reedsy

The same freelance platform also offers a free layout program that works directly through a web browser. However, the quality is not typically great.

- Calibre

This program converts word docs to mobi files for eBooks or other print-ready files. It's free to use, but not exactly user-friendly.

- Vellum

The best program available for an independent author to DIY this step. But, the license to the program costs $250 upfront and is only available for iOS.

PUBLISHING PLATFORMS

- Kindle Direct Publishing (KDP)

KDP distributes directly through Amazon, and so their platform has arguably the largest hold of the market. They currently offer eBook, paperback, and audiobook production through ACX. Currently, their hardback option is still in beta mode and you cannot make a bulk order of physical copies.

- IngramSpark

Probably second in size next to KDP and offers eBook, paperback, and hardbacks in every specification. You can publish with either POD or make bulks orders for only the cost of print and shipping. This platform is best to get physical ARC copies of your book.

- iBooks

Apple's digital publishing platform that represents a small portion of the market.

• Barnes & Noble Press

A very small corner of the market that only offers digital publication.

FOR FOR FINDING SPEAKING ENGAGEMENTS

• Eventbrite

A well-known platform to find general events. You can specify your search and reach out to find local opportunities.

• National Speakers Association

While there is a membership fee and required dues, this does seem like a helpful organization for those who are looking to improve their current skills and develop a full speaking career.

. . .

- Conferize

Another event platform where you can find any event based on type or location, or create and manage your own.

www.ingramcontent.com/pod-product-compliance
Lightning Source LLC
Chambersburg PA
CBHW051857160426
43209CB00039B/1973/J